Saline District Library
W9-BUB-601

Kindness
IN A CRUEL WORLD

NIGEL
BARBER

Kindness IN A CRUEL WORLD

THE EVOLUTION OF *Altruism*

Prometheus Books
59 John Glenn Drive
Amherst, New York 14228-2197

Published 2004 by Prometheus Books

Inquiries should be addressed to
Prometheus Books
59 John Glenn Drive
Amherst, New York 14228–2197
VOICE: 716–691–0133, ext. 207
FAX: 716–564–2711
WWW.PROMETHEUSBOOKS.COM

08 07 06 05 04   5 4 3 2 1

Library of Congress Cataloging-in-Publication Data

Barber, Nigel, 1955–
Kindness in a cruel world : the evolution of altruism / Nigel Barber.
p. cm.
Includes bibliographical references and index.
ISBN 1-59102-228-2 (hardcover: alk. paper)
1. Altruism. 2. Altruistic behavior in animals. 3. Evolution (Biology)
4. Evolutionary psychology. I. Title.

BF637.H4B375 2004
155.2′32—dc22

2004011156

Printed in the United States of America on acid-free paper

Contents

PART 3: THE SOCIAL IMPACT OF KINDNESS

PART 4: KINDNESS AND POLITICS

Acknowledgments

I am grateful for the help received from many people at various stages in the conception, planning, and writing of this book. My editor at Prometheus Books, Linda Regan, influenced the content of the book and provided many helpful and thoughtful comments on the manuscript. My research was aided by the work of the Interlibrary Loan Department at the Portland Public Library, Maine, including Rita Gorham, Eileen MacAdam, and Anne Ball. Schadijah Camp of the Library of Congress helped with illustrations.

Trudy Callaghan provided informal editorial assistance. I am grateful for her counsel and support. David Barber-Callaghan also helped by providing clever analogies.

I am thankful to many friends and colleagues in research whose important ideas and findings in the evolutionary study of altruism put evolutionary psychology on the intellectual map. Many are, or were, members of the Human Behavior and Evolution Society. I also owe a debt of gratitude to the online community of evolutionary psychologists whose challenging ideas promote intellectual fitness. We all mourn the passing of Linda Mealey, who contributed generously to my previous book, *The Science of Romance*, during her final illness.

Introduction

Making the case that people are naturally helpful to others seems a hard sell in a world preoccupied with global terrorism, corporate swindlers, and pedophile priests. Yet, none of these manifestations of evil minimizes the altruistic motive that springs eternal in the human breast. Kindness exists, but it struggles to stay afloat on an ocean of cruelty that is the default condition for organisms competing for existence on this planet.

This book describes actions rather than philosophies. It is also more concerned with concrete examples of helpful behavior than with the computer models many scholars use to decode them. Altruism is defined as actions that help another individual at some cost to the altruist. The main advantage of this approach is that it allows us to discuss human altruism in terms of the evolutionary ideas that account for altruism in other species.

The biological definition of altruism is sometimes criticized as defining altruism out of existence by implying that apparent acts of selflessness are really only selfishness in disguise. The argument goes that such acts are undertaken to (1) make the altruist feel better, (2) increase the reproductive success of the altruist, or (3) increase the prevalence of genes for altruistic behavior. Yet, altruism *is* real, in the sense that it is predicated on evolved moral emotions like empathy and shame. Moreover, it is possible for humans and other species to engage in altruistic actions that have no ulterior motive, except whatever pleasure comes from the action itself, and no delayed benefit of any kind. (The only requirement for altruistic tendencies to evolve is that they should *generally* increase the biological success of individuals expressing them, not that every altruistic act should have a reproductive payoff.)

There is nothing pie-eyed about the case for altruism as an animal adaptation. The simplest, and most prevalent, type is parental care that happens to be well developed in female mammals, some male fishes, and birds of both sexes. Such altruism is easily explained in terms of kin selection, or the promotion of one's genes through reproduction by close relatives. It is predictable, obligatory, and thus not very interesting, except when it fails, as in the case of parents who abuse or abandon their children.

The farther one moves from nuclear families, the more interesting, and more fragile, altruism becomes. Solitary altruists cannot exist in a group of nonaltruists. Altruism is thus a property of groups, or social contexts, as much as it is a property of individuals. Our ancestors survived the uncertainty of hunting large animals by pooling their kill in a prehistoric insurance system that was to the benefit of all. Relying on meat alone, the average solitary hunter might be successful in bringing home a large game animal once a month, which is far too infrequent to enable survival. It makes more sense to share the meat while it is fresh, rather than leaving it to rot in a few days, because sharing entitles the donor to a steady supply of fresh food from other kills.

With some excitement, animal behaviorists discovered a similar form of cooperation among vampire bats. These bats can survive for only a few days if deprived of a blood meal. Sated animals regurgitate food to sustain their famished friends. Such altruism is remarkable among animals that are not close relatives, but it is not automatic, or unconditional. If a bat refuses to share with her hungry friend, she is liable to be kicked out of their shared roost site and may starve. Our human ancestors were similarly protected against cheats by emotions such as anger and indignation that helped keep other members of their community in line. Without such checks and balances, any altruistic relationship among nonrelatives would inevitably collapse as cheats prospered at the expense of cooperators.

Food-sharing altruism promoted survival among our hunter-gatherer ancestors, but the sense of community of interest clearly did not stop at the cooking fire: it extended to cooperative childcare, medicine, entertainment, and so on. Cooperative behavior and the many psychological adaptations necessary to maintain it did not go away when our ancestors shifted to agricultural food production. In settled communities, people helped each other to survive the uncertainties of agricultural production, with its droughts, fires, damaging storms, and pests.

In the earliest urban civilizations, people cooperated to build colossal irrigation systems and cereal storage facilities that finally took much of the uncertainty out of agriculture. In the modern world of highly centralized governments, the initial impulse for centralization came from military threat, but large government is maintained by altruistic social programs, including welfare, education, and security for seniors. Adaptations for altruism that arose from a hunter-gatherer way of life can thus be expressed in very different ecologies. Even so, some aspects of urban living, particularly anonymity, decrease the level of interpersonal altruism and increase crime.

There is a qualitative difference between everyday food sharing, for example, and the more costly forms of heroic self-sacrifice, such as death while fighting in a war. Sharing lunch is one thing, giving up one's life is another. Yet, this book makes the case that both types of altruism are founded on the same adaptations for mutual support. Many of the combatants in wars have not chosen to be present, of course. They were pushed into service by their governments, their tribes, their families, or, even, by the desire to impress their girlfriends. In many cases, there has been wholesale manipulation based on social class and ethnicity, with those on the bottom bearing the brunt of military service while the elite either bribe their way out of combat situations entirely, or purchase military rank that keeps them out of the most dangerous engagements. In the Vietnam War, for example, more African Americans from southern states served than would be predicted from their proportion of the population.

Despite such evidence of manipulation, military service *is* altruistic, in the sense that the combatants sacrifice their personal welfare for the good of others. Throughout recorded history, families have been willing to send their young men into battle to preserve a power structure that favored the survivors. In return for their risky service, surviving soldiers used to be rewarded with high social status that made them more attractive to women. There was thus some potential for personal gain, although hardly enough to account for the great risks undertaken.

One of the toughest conclusions of this book is that in-group altruism can translate into out-group aggression. A soldier supports his own nation by killing the sons of an enemy nation. This irony is all the more remarkable given that group allegiances can be so easily, and so arbitrarily, formed. Just about any distinguishing feature of a group, such as wearing red buttons rather than blue, can promote in-group identification and social hostility against other groups, according to experiments by social psychologists. Such instant groupishness means that we are very good at transcending our groups of origin and assimilating in new ones, whether they are commercial companies, new countries, fan clubs, or religions. Ultimately, such flexibility will be important in the forging of a world community in which nations can cooperate to counteract global terrorism and solve environmental problems, fight AIDS, and confront hunger and poverty.

Even for the most fundamental kinds of groups, such as households or local communities, effective cooperation requires that individuals behave in predictable ways. In the jargon of social psychology, they conform with norms of expected behavior. Such conformity is a form of altruism because the individual sacrifices immediate satisfaction of impulses in the name of predictability and preservation of social harmony. Imagine a world in which people drive on whichever side of the road they like. A small sacrifice of self-determination in this matter is beneficial to all users of the road. In principle, this is a win-win situation little different from the rational food sharing of our subsistence ancestors. This book suggests that it is also based on adaptations of the human brain for cooperative behavior.

We cannot understand group cooperation without understanding the conflict and competition that threaten it. Economists and students of altruism are preoccupied with the instability of altruistic systems in the face of individual selfishness. When herders share common pastures, there is a danger of overgrazing, leading to soil erosion, invasion of weeds, and destruction of the fragile grassland ecology. This scenario is referred to as the *tragedy of the commons* and has been used to account for many environmental problems, such as smog and global warming. The key issue is that the benefits of pollution, such as those from driving a large, powerful vehicle, go to the polluter alone, whereas the costs to the global environment are shared by everyone. Economists tend to be pessimists, and many believe that we are in the midst of a global tragedy of the commons in which individual, and national, selfishness stretches the earth's fragile ecologies past the point of no return. I do not share that pessimism, believing that as the population explosion subsides, the global community can agree on self-imposed restraints for each nation that arrest greenhouse gas emissions and preserve forests. Most people want to inhabit a global community that allows the same level of cooperation among nations as among the members of a hunter-gatherer community.

Arguing that human beings have an evolved propensity for cooperation does not mean that they will always behave in cooperative ways, even within their own communities. The most spectacular failure of altruism involves violent criminals. The drive-by robber who kills a convenience store clerk to obtain a hundred dollars and a carton of cigarettes is often held up as an example of innate human depravity. Unfortunately, some individuals are indeed born without the capacity to develop a conscience. (Others fail to develop sensitivity to others because of the brutalizing conditions of their childhood.) Antisocial personalities, formerly referred to as moral imbeciles, visit misery on themselves and everyone connected with them. Fortunately, they are also rare. Most of us do have a well-developed moral intelligence.

Violent criminals are the exception that tests the rule. Another striking exception is the unusual case of mothers who kill their children, thus violating our expectations of maternal altruism. Ironically, in at least a few cases, mothers kill their children with the misguided intention of sparing them from suffering. This may happen if the mother is suicidal and plans to take her own life. Homicidal mothers are often mentally ill and rarely have anything to gain from the death of their children.

The same can hardly be said of white-collar criminals, such as those who recently defrauded American investors of billions of dollars. One of the most striking aspects of such criminals is the moral grayness of the world they inhabit. Some of the key players in the Enron debacle vehemently deny that they did anything illegal. Whether this is true is a matter for the courts, but experts in corporate law are quite divided. As a moral issue it is clear that their unfettered greed drove a once-profitable company into bankruptcy, resulting in huge job losses and the destruction of employee retirement savings. Crooked executives wound up with a lot of money in their pockets, and hapless employees and investors had none.

Although money allows us to engage in cooperative transactions with people around the world, it can stimulate competition and undermine altruism. This fact emerges from comparisons of altruistic motives in children around the world. In poor countries, youngsters are generally much more concerned with the welfare of others than is true of wealthy countries like our own. The reason is simple: much is asked of them. Children living in extended families in agricultural communities take care of younger siblings and help aged relatives. They also do a lot of the work on family farms. In contrast, children in affluent countries are asked for very little and receive much. They develop a strong sense of entitlement but very little sense of their obligations to others.

As countries become economically developed, the strong ties among members of local communities are weakened by the transition to a monetary economy. This pattern was studied among the Igbo tribe in Nigeria, where voluntary help on farms, usually by kin, was replaced by wage labor. The problem with monetary transactions is that they promote cooperation in temporary relationships among strangers and undermine the system of mutual obligation that was practiced since time immemorial and allowed individuals to make substantial voluntary contributions to their communities, rewarded mainly by affection and respect, as well as the security of knowing that there was a fund of altruism to draw from in time of need.

The moral of the Igbo story is that altruism is comparable to physical fitness. We cannot expect children to become athletes without any opportunity for physical exercise. Neither can we expect them to help others if they receive no training in altruism. Our evolutionary history has equipped us to develop skill and strength through training. It has also provided us with altruistic motives that grow stronger from exercise.

PART 1

Altruism in Man and Beast

CHAPTER 1

Altruism

BIRDS DO IT, BEES DO IT, PEOPLE DO IT

The idea of altruism can be a very slippery one. If a person gives blood only to make himself feel superior to nondonors, is he really behaving altruistically? Philosophers will puzzle over such matters until the end of time. Biologists have a much simpler definition of altruism. In biology, animals struggle to do two things: survive and reproduce. An altruist is one who puts the survival or reproduction of another individual before his own. Worker bees do not reproduce, for example, but they serve as nursemaids, raising the brood of their mother, the queen. Because they put the queen's reproduction before their own, they are altruists. Exactly the same kind of altruism is found among humans. Thus, female Chinese silk workers of the nineteenth century formally refrained from marriage in order to send their wages back to support their families at home. These members of the "marriage resistance" who chose never to marry and never to have sex with men were thus as sterile as worker bees.

ALTRUISM IN SOCIAL INSECTS

At present, many people are very excited about the insights that evolutionary psychology provides about everyday phenomena from physical attractiveness to violent crime to fashions of dress and sexual behavior. It is hard to imagine that all of this began with the study of reproductive altruism in bees and other social insects.

How could the capacity to forgo reproduction be built into sterile social insect workers by nature? The modern Darwinian answer to this problem was simple but profound. Think of reproduction as a process by which copies of genes are propagated into future generations. Worker bees do not have offspring but they devote themselves to caring for sisters who share many of their genes. Within this framework, known as "gene selection," it is irrelevant whether the copies are made directly by individual reproduction or indirectly through the reproduction of a close relative, the queen bee, mother of all the workers.

In the terminology of biologists, worker bees are "altruistic" because they put the reproductive interests of the queen ahead of their own personal reproduction. A great deal of ink has been spilled in trying to understand whether such altruism has anything to do with human self-sacrifice. There are many differences, of course, the most obvious of which is that human beings can reflect on their own actions whereas bees are essentially stimulus-response machines. Yet, the thread of connection is strong enough to make a case for evolutionary ethics. Before moving on to complex examples of human altruism, it is useful to know how altruism could evolve in bees. This was a major riddle for biologists well into the second half of the twentieth century. It flummoxed no less an evolutionist than Charles Darwin (see fig. 1). Before demonstrating why Darwin's theory could not account for the altruism of the worker bees, we need to describe his theory of evolution by natural selection.

DARWIN'S THEORY OF EVOLUTION BY NATURAL SELECTION

Darwin's theory of evolution is simplicity itself, but it is not for the faint of heart. His theory emphasizes the struggle to survive and reproduce in a difficult world where there are no second chances. One might imagine that this dark vision of nature as epitomized by

Fig. 1. Charles Darwin was baffled by altruism among bees (see text for explanation). (Library of Congress)

blood, teeth, and claws was inspired by study of predator–prey interactions, but a formative influence was Thomas Malthus's (1766–1834) *Essay on Population*.[1] It laid out a dismal scenario about how human populations always grow more rapidly than increases in the food supply, so that the population expands steadily until it is checked by famine and disease. Whether this scenario is accurate is beside the point here. What is important is that it contained within it the germ of Darwin's concept of natural selection. We are on fairly firm ground in giving Malthus some credit because at least one other person who cracked the problem of evolution was Alfred Russell Wallace (1823–1913), who had also read Malthus. This was quite unusual reading because Malthus was deeply detested for conveying such a pessimistic message in an age of optimism. (Recent scholarship, by Prof. Paul Pearson of the University of Cardiff, concludes that true priority for the theory of evolution goes to James Hutton [1794], who was sixty-five years ahead of Darwin [1859] and who could not have been influenced by Malthus's ideas on population because they were not published until four years later.)[2]

If there is a struggle to survive among human populations, the same must clearly be true of animals (not to mention plants) that inhabit the natural environment. Here there is little in the way of a cushion against adverse conditions like periodic food shortages, and the battle against predators never ends. Darwin realized that some individuals are slightly better suited to the way that they make their living than are others. Imagine a population of leaf-eating deer inhabiting an African savanna. A severe drought takes hold. Most of the shrubs die except for the large ones with deep root systems that can still reach the water table. Many of the leaf-eating deer perish from starvation. Survivors have one thing in common: most have unusually long necks that enable them to feed from the tops of large bushes where the few remaining leaves are to be found.

Darwin would say that the long-necked individuals are favored in the struggle for existence.[3] The key component of Darwin's theory is not really survival, however, it is reproduction. When the long-necked survivors breed, their offspring will have longer necks, on average, than earlier generations because the offspring resemble their parents. Extended over thousands of generations, the ordinary leaf-browsing deer get turned into giraffes whose body design is the animal equivalent of a cherry picker. Giraffes, in other words, are exquisitely designed for the way that they can survive, foraging fourteen feet into the air. Evolution by natural selection can thus explain why animals are so well adapted to their ecological niches. Darwin's theory solves the problem of adaptation and it does so in a completely mechanical way, without any need for divine intervention, as the natural theologists of his day, such as William Paley, had argued.

The term "natural selection" deserves some further comment. Given that Darwin knew nothing about genetics, how could he be so confident that the offspring would resemble the parents and that some trait, like the giraffe's neck, could be progressively modified over many generations? It turns out that like many other English gentlemen of his day, Darwin had a keen interest in animal breeding, as well as the development of good strains of agricultural crops. He was well aware that horses could be selectively bred for speed. He even indulged a passion for breeding show pigeons. Each of these pursuits clearly showed that animal breeders could manipulate some desired trait by altering the heredity of their animals through selective breeding.

Horse breeders mate the fastest stallions to the fastest mares and hope that the offspring will be capable of winning races. This has resulted in sharp increases in the speed of English racehorses over the past couple of centuries, although some experts believe that racehorses of today are as fast as they will ever be.

Darwin's central idea of natural selection can be compared to animal breeding, or artificial selection. In Darwin's theory of evolution, the natural environment fulfills the role of the animal breeder, "deciding" which individuals will survive and reproduce. If artificial selection could alter the hereditary features of animals, then Darwin was on fairly solid ground in assuming that natural selection could do the same. The natural environment selectively breeds animals that are suited to surviving and reproducing in that environment. Stated in this way, it is clear that doubting natural selection is as unreasonable as doubting that artificial selection can occur.

Darwin's theory of evolution is so simple that you could write it on a postage stamp. Yet the implications are overwhelmingly complex, particularly when applied to our own species. Before doing that, it is necessary to revisit Darwin's bee problem, the problem of reproductive altruism.

REPRODUCTIVE ALTRUISM

One of the critical insights of evolutionary theory is that there is a competitive struggle to survive and reproduce. This struggle is played out very literally in the animal world in terms of competition for food, for safe places to evade predators, and over access to mates. Individuals that are successful in this struggle survive and reproduce. Those that fare less well die early, leaving no offspring. These facts boil down to the harshly competitive interactions we see in all species when a vital resource of some kind is at stake. They are at odds with the altruism of social insects where workers serve the reproductive interests of their sister, the queen, and never reproduce themselves.

It is not difficult to demonstrate competition and selfishness in the animal world. Primatologists used to provide food for animals they studied, making them easier to locate and approach. They soon noticed that animals became highly aggressive toward other members of the species around provisioning sites and were forced to discontinue the practice both because the animals injured each other and because the artificial introduction of large amounts of "free" food so distorted the natural behavior of the animals that it completely undermined the purpose of the research, that is, capturing the behavior of the animals in their natural environment.

Selfishness is favored by natural selection so that it should be present wherever an animal behaviorist cares to look. In general, that is what they find. Selfishness is expected because selfish individuals would generally compete more successfully for resources and leave more offspring. Altruism is unexpected, and that is why it has exerted such a fascination for biologists from Darwin onward. How can altruism emerge in a world of Darwinian selfishness?

That simple question lay unresolved until the early 1960s, when

English biologist William Hamilton offered an elegant, and compelling, solution to the enigma.[4] The key issue is that altruism can arise by natural selection if it is directed toward close relatives of the altruist.

This point is better appreciated by imagining that an "altruistic" gene arises that makes its owner behave kindly to all other members of the species. New genes crop up naturally through copying errors, known as mutations. It is unlikely that a single gene could have this effect, of course, but a dozen, or a hundred, might. The logic is the same: for the purposes of the thought experiment, we wait until all of the necessary genes are carried by an individual, making him really nice to others. He defers to others when there is a conflict over sexual access to a female in heat. If it is a female, she cares for unrelated offspring as much as her own.

There is a very simple reason why such an altruistic gene cannot arise through natural selection. Its owner would get taken advantage of. If there were a food scarcity, the altruistic individual would die of hunger. He would lose out in the competition over mates. He would neglect his own offspring in favor of the young of others. Clearly, such an altruistic individual would have no surviving offspring. The altruistic gene would die out in its first generation.

Yet, there is a tremendous amount of altruism in the natural world, the most striking and obvious of which is the self-sacrifice of parents for their young. The lioness not only feeds her cubs with milk produced in her mammary glands, but also is prepared to risk her own life in their defense.

Self-sacrifice is carried to an extreme among some animals.

Altruistic suicide occurs in a number of insect species. Honeybees (see fig. 2) will give up their lives in defense of the hive, for example. When they sting an invader, their stingers are ripped out of their abdomens and they soon die of the wounds. Clearly, bees do not consider their actions in the way that humans would, and cannot know that their defense of the hive will result in their own deaths. This is referred to as suicide because the individuals kill themselves, not because they intend to do so.

Suicide also occurs among Australian social spider mothers (*Diaea ergandros*) in an even more grisly way: they are eaten by their own

Fig. 2. Honeybees, long used as an analogy to complex human societies: individuals give up their lives in defense of the hive. (Library of Congress)

young.[5] Before death, mothers stock their ovaries with eggs that cannot develop but will enhance their nutritional contribution to the young when they make a meal of Mom. Human suicide, during war or in the name of a political cause, has little to do with genetic relatedness as such but provides an eerie counterpoint to suicide in social insects.

Apart from their intrinsic fascination, such examples suggest that a high level of altruism is possible among close relatives like parents and offspring. Why? That is where Hamilton's gene selection plays a role.

We have seen that indiscriminate altruism cannot work in a world of selfish individuals because the altruist will get taken advantage of and suffer an extreme loss of fitness (i.e., reproductive success). In gene selection, the focus shifts from the point of view of an individual producing young to the rather more abstract notion of genes being copied into subsequent generations. The reason for this abstraction quickly becomes obvious. I can transmit my genes into the next generation by fathering children of my own but the same goal can be accomplished through reproduction by close relatives. Having many nieces and nephews, for example, is as effective a way of propagating genes into the next generation as having two children of my own. From the gene's perspective, it does not matter whether the copy is made by direct reproduction or indirectly through the reproduction of blood relatives.

To an evolutionist, fitness means reproductive success. Before Hamilton's gene-level selection, reproductive success was measured in terms of the number of offspring produced. Thus, a person producing two offspring would have copied his genetic material once into the next generation (50 percent going to each child). Our children are special because they are so closely related to us, but they are not the only ones to share a large portion of our genotype (i.e., our individual genetic makeup). A person who has no children of his own but eight nieces and nephews is transmitting two copies of their genes to the next generation, given that each niece or nephew shares a quarter of a person's genes. Without doing anything, they have the same inclusive fitness (i.e., copies of their genes going to the next generation) as the person who was kept up at night by two howling babies! The term "inclusive fitness" describes copies of our genes that are transmitted into the next generation via reproduction by close relatives as well as by direct reproduction.

The concept of inclusive fitness helps us to understand why genes that underlie altruism can be maintained by natural selection if the altruism is discriminating, or directed at an appropriate target, namely a close relative. When we talk about genes as altruistic, this is intended only in a metaphorical sense as contributing to altruistic behavior. Strictly speaking, genes can do only one thing: organize the assembly of proteins that compose the building blocks of animal life. Genes are thus rather like the blueprint of a building. The blueprint does not evoke emotion as a splendid building can. Similarly, genes are neither selfish nor altruistic in themselves but they organize the construction of a brain and a body that is capable of feeling emotions and being swayed by moral sentiments.

The simplest case of altruism is parental care. Not all multicellular animals care for their young. Many devote their efforts to producing large numbers of offspring, the great majority of which will die before reaching maturity. Thus, female oysters may lay millions of eggs but their parental investment ceases when the eggs are laid. In fact, there is nothing to prevent them from accidentally feeding on their own young that teem in the waters around them, and they do.

Animals that produce fewer offspring typically invest more in each of them. This pattern is marked in birds and mammals that already invest a great deal of bodily energy in the offspring by the time they hatch or are born. Such investment is obligatory, or fixed. Investment after birth, or hatching, is much more variable. Thus, most reptiles spend little time caring for their offspring while most birds lavish extensive parental care on theirs. Even among birds, parental care is highly variable, particularly for males, some of whom are equal partners with the females in caring for nestlings and some of whom do almost nothing.[6]

Parental care evolves in a context in which parents have already invested a great deal in the young by the time they emerge from egg or womb. Further investment, in the form of parental care, is warranted if it increases the chances that the young will survive and reproduce. Parental care is a particular type of altruism, but from the point of view of its evolutionary origin it is no different from others.

Suppose that there is a species of birds in which all of the labor of feeding the nestlings is performed by the female. Now, suppose that a mutation (or series of mutations) arises that makes males more likely to help out around the nest. If their help makes a difference, more of the young will reach maturity. Half of the male survivors will inherit the altruistic paternal-care gene and half of the female offspring will transmit it to their offspring. If nestlings continue to do well following paternal care, the altruistic gene will quickly spread throughout the population, beating out rival genes, or alleles, that promote male laziness.

When you think of parental care as a kind of altruism, and

clearly the most widespread type, then the whole question of the evolution of altruism becomes quite clear. Parents care for their young because doing so increases their reproductive success. In the process, it increases the frequency of genes for parental care of offspring that are carried by those young.

Note that parental altruism can be maintained by natural selection because it is discriminating. Parent birds do not care for fledglings at random, but only those hatching in their own nest. Admittedly, the system is not perfect. Cuckoos and cowbirds are adept at inserting their own eggs into the nests of other species. The young cuckoo hatches early and tips the other eggs out of the nest to their doom so that all of the parental feeding efforts go to the invader. Even this evident failure of parental altruism may illustrate adaptive design relevant to parental discrimination. When parent birds feed their young, they are stimulated to do so by a brightly colored display created by the gaping mouths of the hungry youngsters. This display is species typical. Parent birds are stimulated to feed the cuckoo because the pattern of its gape display has evolved to match that of the host species. Even in its most conspicuous failure, the parental investment system of birds shows evidence of design for discrimination. There is a coded message in the gape display, but the code has been broken by an enemy![7]

The discriminating nature of parental investment is illustrated in many other ways. During the fall, ewes try to wean their lambs. The lambs persist in their suckling attempts. Eventually, the irritated mother will buck at the lamb, or even kick it in the head when it attempts to steal her milk. Eventually, the lambs are forced to feed themselves by grazing and the mother stops lactating for that year. Conflict over weaning is theoretically interesting because it is in the reproductive interest of the mother to wean early so that she can gain weight, put on fat stores, come into heat, and conceive next year's lamb. Conversely, the lamb is better served by late weaning because the additional maternal investment is worth its full value to the individual as opposed to contributing 50 percent to inclusive fitness by going to a sibling that shares half of the individual's genes.[8]

An even more obvious example of discrimination in parental

care is the fact that parents invest a great deal in children but children are far less likely to invest in their parents. This asymmetry is perfectly logical. Parents invest in their dependent children so that those children can survive and reproduce, thereby contributing to the inclusive fitness of the parents—transmitting their genes to future generations. Mature offspring could also, in theory, invest in their parents and thereby increase their own inclusive fitness by raising more siblings, but they are unlikely to do so. It makes more sense to invest in their own young while they have an entire reproductive career ahead of them rather than that of their parents, whose reproduction is close to the end.

Young do occasionally help their parents to reproduce, however. Among birds, it is referred to as the helpers-at-the-nest phenomenon. Species whose subadults help to raise their parents' next clutch include: green hoopoes in Kenya, the African white-fronted bee eaters, African pied kingfishers, Florida scrub jays, and the endangered red-cockaded woodpeckers of the United States. In general, birds help at the nest only when it is difficult, or impossible, for them to reproduce successfully themselves. This is clearly true of the woodpeckers. The comparative rarity of helpers at the nest is quite difficult to explain from a gene-selectionist perspective, but it seems that investment in "young reproductives"—themselves—is generally a more viable strategy than investment in "old reproductives"—the parents.

Difficulty in breeding independently for the red-cockaded woodpecker is due to scarcity of suitable habitat. These birds inhabit stands of mature living pine in southern US forests. They nest in trees infected with "red heart," a kind of fungal disease that softens the wood at the center of the tree, allowing them to dig out nest cavities and roosting spaces. Limited habitat means that landowners who have red-cockaded woodpeckers on their land are legally prohibited from cutting down trees.[9]

Females of this species disperse at maturity and males stick around to claim the ancestral breeding territories. Instead of breeding themselves, in direct competition with their parents, the younger birds postpone reproduction and help to care for younger siblings. Researchers have observed that nests with helpers raise more young than those without. The usefulness of male helpers may explain why there are almost twice as many males as females, even though most bird species have the same number of each sex. Among their other peculiarities, these woodpeckers have the capacity to alter the proportion of males to females during their reproductive careers. A young mother produces almost all sons because males are more likely to serve as helpers as they stick around to inherit valuable nest territories. After she has produced enough helpers to raise her offspring, she starts producing more daughters.[10]

Such helping is not peculiar to birds, although it is more common for them because of their intense parental investment in nestlings. Among mammals, much of the parental care is provided by the mother, who must be present to suckle the offspring. Providing solid food that must be carried to the young is comparatively unusual, except among predators. African hunting dogs are a case in point. Males stay on with the parents after maturity and help them by feeding, and protecting, the next litter of pups. Hunting dogs range over large distances and carry meat back to the pups—in their stomachs! The food is regurgitated to provide meals for the dependent young.

Like the red-cockaded woodpecker, African hunting dogs produce more males, reflecting the important contribution of male helpers. Thus, male helpers carry as much meat to the pups as the father does.

Human helpers are so common that it scarcely seems worth mentioning. Children everywhere are expected to perform some childcare services for their younger siblings. The phenomenon is found to an extreme degree among Hungarian Roma (often referred to as Gypsies), where young women often stay on to help their mothers, even after they themselves have married.[11]

HUMAN HELPERS AT THE NEST

Some degree of sibling help with child raising is found in many hunter-gatherer societies. Such help is not always an unmixed blessing because of rivalry between siblings. Thus !Kung children left in charge of younger siblings may abuse them. In rare cases they even attempt to drown them. This means that young helpers have to be supervised carefully. In general, though, it is reasonable to assume that the benefits to parents of human helpers at the nest greatly exceed their costs.

Apart from recruiting siblings to provide free baby-sitting services, most hunter-gatherer societies enjoy a cooperative child-rearing system in which adults other than the parents hold infants at least some of the time, giving the mother the opportunity to gather food unencumbered. Such cooperation is found at an extreme among the Efe pygmies in the Ituri Forest of Central Africa (and is by no means unique to their society). Small babies receive a great deal of their care from individuals other than the mother, including fathers, who hold the infants for about 20 percent of the time. At the age of three weeks, infants spend 39 percent of their time in physical contact with nonmaternal caregivers, and this increases to 60 percent at eighteen weeks of age.[12]

There are many possible reasons why there is so much cooperative childcare among the Efe. One is that they have an usually high degree of infertility among women of reproductive age, and infertile women are more likely to help than women with children of their own. This is hardly surprising because they would have more time. Efe women are infertile because of reduced ovarian function and low progesterone production. This is due partly to their very active lifestyles—they frequently carry heavy loads for many hours—and partly due to the prevalence of many diseases including malaria, gonorrhea, infections, and internal parasites.

Older women, whose reproductive careers are over, play a more important role as helpers than women of reproductive age. Young girls spend more time caring for infants than young boys, which makes sense if they are learning how to be more effective mothers for their own future children.

The most important determinant of which Efe youngsters will help parents is their degree of genetic relatedness to the infant. Efe helpers thus promote their own inclusive fitness (or gene transmission via relatives) by contributing to the survival and eventual reproductive success of close relatives. This kind of altruism can therefore be seen as a type of kin-selected altruism. Instead of investing in their own children, or direct reproduction, helpers copy their genes into future generations indirectly through the reproduction of blood relatives.

So far so good. Up to this point, genes have been referred to in a loose, even metaphorical way. We do not really know which genes are involved, or what their mechanisms of action are. Such gaps, which may or may not be filled by future scientists, do not take away from the logical force of the kin selection argument to any degree. Kin-selection theory claims that altruism can evolve if it promotes inclusive fitness. It allows us to make predictions about how people may behave and therefore provides a useful way of organizing facts and building our knowledge about human social behavior.

Yet, kin selection is obviously very far from being the whole story. Thus, it does not explain the relative absence of cooperative childcare in many societies, including our own. Isolation of nuclear families in modern cities would make it quite impractical for close relatives to care for each other's children on a regular basis. American women differ little, if at all, in the genes affecting their altruistic feelings toward dependent infants. Those nurturant impulses get expressed differently in urban North America compared to the Ituri Forest in the Democratic Republic of the Congo. Evolutionary explanation has as much to do with how people cope with environmental variation as with genetic influences on behavior.

What do human helpers have to do with the problem of sterile insect castes that were so bothersome to Darwin? On the surface, the resemblance might seem far-fetched. At the level of gene selection, it is identical.

DARWIN'S BEE PROBLEM REVISITED

Darwin was baffled by nonreproducing worker bees, but he offered the best explanation he could think of. He imagined that natural selection must act on the entire bee colony. Just as a complex animal is made of many organs, Darwin thought of a colony of social insects as a sort of superorganism in which different functions were divided up among the various castes for their mutual benefit. The queen, for example, was responsible for the production of eggs and was thus similar to the ovaries in a mammal.

Ingenious though it was, the superorganism idea was based on a crude analogy that doesn't work. After all, each individual worker bee is independent and complete and there is no obvious or compelling reason why they could not reproduce independently. There must be some good reason why whole castes of workers forgo reproduction, not just among honeybees but also among other hymenopterans, an order that includes social insects such as ants and wasps.

Gene selection provides just such a compelling reason, as first pointed out by English biologist William Hamilton in theoretical papers published in 1964.[13] His explanation hinges on the rather strange reproductive system of honeybees and some other social insects. Fertilized eggs turn into females who have a pair of each chromosome, just like people, and are thus said to be "diploid" (or doubled). Eggs that are not fertilized develop into males that possess only one of each chromosome and are thus referred to as "haploid" (or "monoploid," i.e., single). This odd method of sex determination goes by the unlovely name of haploidydiploidy.

Hamilton developed the gene-selectionist implications of this reproductive system by analyzing the degree of relatedness between offspring in a social insect colony and the queen on the one hand and the workers on the other. As in people, the queen's fertilized egg combines one chromosome from the male and one from the female to make up each chromosome pair found in a diploid cell. Like a human mother, her degree of relatedness to the young is 50 percent.

The degree of relatedness among workers is calculated as follows. They receive the same set of chromosomes from the father (assuming that the mother mates only once) because he has only one set to give. This means that they are 50 percent related from the paternal line. On the mother's side, they share half the genes they receive, on average (based on getting a random half of the mother's paired chromosomes). The mother thus contributes 25 percent genetic similarity to offspring, just as is true of humans (who receive the other 25 percent similarity from the father in the case of full siblings). Adding the 50 percent genetic similarity from the father with the 25 percent similarity from the mother means that worker bees are 75 percent related to each other. This means that they are halfway between full siblings and being genetically identical clones.

Hamilton recognized that the remarkably high degree of relatedness among worker bees, and other sterile hymenopteran workers, had some profound implications for their altruistic behavior toward sisters. First, the apparent reproductive sacrifice of letting the queen do all the reproducing was no sacrifice at all from the perspective of gene selection, or inclusive fitness. The workers were actually more closely related to their sister workers than to the queen, their mother. In this Alice-in-Wonderland scenario, the workers contribute more to their inclusive fitness by helping their mother to reproduce than they would if they reproduced themselves. It is not difficult to see why genes promoting worker sterility would invade the gene pool, beating out alleles supporting independent reproduction. (Alleles are alternative forms of a gene.) One unresolved problem for Hamilton's theory is that queens frequently mate more than once, which dilutes the level of relatedness among workers to that of ordinary siblings, or less.

Fascinating as the theory is, the mechanism through which workers become sterile is just as strange. Future queens receive a special diet, known as royal jelly, that enables their reproductive systems to become fully mature. Without this special nutrition, the larvae are destined to become sterile workers. Sterility is thus evidently under the control of the workers and a function of their behavior.

Viewed in terms of gene selection, the altruism of worker bees is no different in principle from the altruism of helpers of other kinds, human or nonhuman. The major difference is one of degree. Worker bees are committed entirely to indirect reproduction through their mother.

Biologists have often claimed that the extreme kin-selected altruism found in social insects was due to their peculiar reproductive system (i.e., haplodiploidy), but this is not true.[14] Thus termites have a similar social system to bees but they are diploid. The recent discovery of a similar social system in the naked mole rat, which is diploid, like all other mammals, provides further evidence that a high level of altruism can occur without haploid males. Mole rats live in colonies of about eighty individuals containing a single reproductive female, the queen, and about three reproductive males, all of whom typically mate with the queen whenever she is in heat.[15]

The mechanism of reproductive suppression in mole rats is very different from that of social insects. Female mole rats are actually capable of reproduction and will come into heat rather quickly if removed from their colony. Experiments have shown that odors from the queen are not sufficient to suppress reproduction, a mechanism of reproductive suppression that is common in mice. The queen apparently bullies other females by pushing at them with her snout, and this process of intimidation inhibits their capacity to ovulate. When the queen is in the final stages of pregnancy, she is too heavy to go around intimidating the other females, and one may occasionally come into heat and reproduce.

On its face, reproductive suppression in mole rats would seem to be a matter of simple reproductive competition between the queen and other mole rats, but this is not the complete story. After all, why are the females so absurdly sensitive to a little shoving from their queen? This mystery becomes less perplexing in view of the very high degree of relatedness among members of these inbred colonies. They actually outdo the honeybees with an average coefficient of relatedness of .81, compared to the maximum of .75 for social insects. In the presence of a vigorous breeding female, other females evidently do more for their inclusive fitness by failing to reproduce at all than if they were to swamp the colony with pups that could not be raised.

Hamilton provided a compelling explanation for why extreme kin-selected altruism could evolve in highly social animals like bees and mole rats. Kin selection is most often manifested in the care of dependent young, but that is by no means the only setting where it occurs.

HELPFUL RELATIVES

Kin selection may be useful for decoding the fascinating reproductive systems of bees, but is it really any use when applied to human societies? Anthropologists were initially skeptical of applying kin-selection theory to humans.

For a long time, cultural anthropology had operated under the assumptions that every society is different; that cultural differences must be experienced to be properly understood; and that human societies are different from animal societies because their interactions are largely determined by culture rather than biology. This general point of view is referred to as cultural relativism and is vehemently opposed to the kind of general explanations favored by evolutionary biologists. If every human society is unique and operates under its own culturally determined rules, then it would be a stretch to come up with scientific principles governing all human societies, much less ones that can be generalized from honeybees, or slime molds, to humans.

These objections were sincere. Cultural anthropologists believed that evidence from field research would bolster their conclusions. They were particularly confident on the issue of human kin selection. Having assembled masses of data on the inconsistent use of terminology defining kinship in most societies, they were adamant that the cultural definition of kinship had little to do with genetic relatedness. Many societies use fictive kin terms in which a woman who is referred to as "sister" is not related by blood. In our own society, we have godparents who are expected to assume certain parental responsibilities even if they are not close relatives. If, as anthropologists believed, human kinship is of largely cultural construction, and if people treated others according to these kin classifications, kin selection could not be a force in human societies.

That was what anthropologist Kristen Hawkes, of the University of Washington, thought when she decided to use her data on the Binumarien people of Highland New Guinea to test predictions from kin-selection theory.[16] This tribe subsists largely on sweet potatoes and pigs, and they are noted for being highly warlike. The Binumarien, like most other people around the world, take enormous liberties with the actual closeness of blood relationships in their kin terminology. Thus, a cousin is sometimes addressed as "brother," for example. To Hawkes's intense surprise, she nonetheless discovered that the Binumarien exercised favoritism toward close relatives. They were more altruistic in their treatment of close rather than distant kin and favored actual brothers over cousins. As this finding illustrates, we need to pay attention to what people do more than what they say in this context, and what they do is consistent with the predictions of kin-selection theory.

Now that anthropologists recognize the importance of genetic relatedness as a determinant of altruism between individuals, they have begun to collect information useful for testing kin-selection theory. Consistent with kin-selection theory, people prefer to interact with close relatives rather than other people they know, according to field research among the African Ye'Kwana and other societies.[17]

Humans prefer to associate with kin, living close to blood relatives when possible. Among the Yanomamo of South America, villages become unstable when they get too big, and the segregation of people into two smaller villages provides a test of the prediction that people will opt to live in proximity of blood relatives. That is exactly what anthropologist Napoleon Chagnon found.[18] People in the two new villages were significantly more closely related to each other on average than members of the large unstable village community had been. They chose to live among close relatives.

Consistent with kin-selection theory, humans, like all other species, invest a great deal more in their own offspring than they do in the offspring of others. This is not always true, however. One glaring exception is the case of infertile couples who adopt nonrelatives and raise them exactly as they would their own children. This simple fact would appear to falsify a key prediction of kin-selection theory that people should invest in close relatives. But this view falls into the trap of assuming that human behavior must always be adaptive for an evolutionary interpretation to work. In reality, all that is required for natural selection to preserve some pattern of behavior, or psychological responsiveness, is that *on average* it should increase reproductive success. Some people adopt unrelated children, but most do not. What is more, the probability of adopting a complete nonrelative in the evolutionary past would have been extremely low given that people lived in small groups of closely related individuals.

We might also imagine that adopting children was peculiar to humans, but it is not. Penguins that lose their chick may occasionally adopt an orphaned chick that is unrelated to them. Adoption also occurs among primates. It is particularly common among Japanese macaques and baboons, for example. Interestingly, Japanese macaque orphans are most likely to be adopted by siblings, or cousins.[19] Presumably the same was true of our human ancestors. The fact that they lived in closely related kin groups helps explain why we have no "defense" against raising unrelated infants almost exactly as we would raise our own.

One interesting way of studying kin-selected altruism, or "nepotism" (as in helping relatives), is to examine the pattern of bequests left by deceased persons in their wills. Most of this research has been conducted in the United States, where the great bulk of most estates is left to close relatives, consistent with kin-selection theory.[20] Adopted children receive exactly the same treatment as natural children in legacies, however, suggesting that the bonds between adoptive relatives may be just as strong as those of blood relatives.

Despite the lack of obvious favoritism in wills, there is some evidence that adopted children are not treated exactly like biological children, even though most parents strongly believe in treating adopted children and their own children equitably.[21] Thus, stepchildren receive less financial assistance for college expenses from their families compared to children who have two biological parents. Similarly, South African fathers of the Xhosa tribe spend more money on natural children during the high-school years compared to stepchildren, and spend more time helping natural children with their homework.[22]

Apart from diminished investment in education for stepchildren in different societies, there is some evidence that there is less personal warmth shown in relationships between stepchildren and parents. Thus, American parents spend three hours less per week with stepchildren than they do with biological children. Mothers also spend less money on the food budget if they have a stepchild compared to all natural children. Folktales about vicious stepmothers, such as the story of Cinderella, may have a grain of truth in them, but they are generally a wild exaggeration of the diminished "natural affection" between parents and stepchildren (or adopted children), to use a charming phrase from British wills.

The difficulty that some parents feel in making an emotional connection with stepchildren is not peculiar to urban environments. The same phenomenon is observed among the Hadza hunter-gatherers of Tanzania. Fathers provide considerably less care for stepchildren than they do for their own children, for example. Moreover, Hadza fathers were never observed playing with their stepchildren. Although the Hadza generally believe that stepchildren are as well cared for as biological children, stepfathers themselves admit that their feelings of affection are stronger toward their own children, possibly reflecting inherited similarities in temperament and appearance.[23]

Such departures from equitable treatment between parent and stepchild are observed the world over. Evolutionary psychologists have studied many practical examples of favoritism toward close relatives, thus confirming predictions drawn from kin-selection theory.[24] One intriguing study of polygamous families in the American West found that children developed much closer relationships with full siblings than with half-siblings (who shared a father and lived in the same home, but had different mothers). This closeness persisted into adult life when full siblings were much more likely to celebrate birthdays together, and help each other out by giving money, or providing baby-sitting services, compared to half-siblings.[25] Relatives connected through the maternal line are also emotionally closer than those related through the paternal line.

An adopted child is more vulnerable to abuse and even homicide at the hands of a stepparent, compared to a genetic parent, suggesting that in the small minority of homes where such problems occur, the mere fact that a child is unrelated could make him more vulnerable to stepparental violence. Other factors might include how familiar the stepparent is with the child and whether he or she has assisted in raising her. The logic is that people who nurture a child from early in her development are much less likely to do anything that will hurt her. There are other possible complications as well. Stepfathers may resent their stepchildren because of jealous competition for the mother's attention and affection, for example—a problem that is more pronounced in stepfamilies than biological ones because children have already built up a relationship with the mother prior to meeting the stepfather.[26]

In investigating the connection between family violence and stepparents, some of the anthropological research is brutally clear. There are several South American tribes, such as the Ache of Paraguay and the Yanomamo of Venezuela, in which a woman's new husband may either personally kill her young dependent children or insist that she kill them herself. The logic is very clear. A new husband does not want the wife to waste time on the progeny of another man that could be spent breeding for him.[27]

Money and property are interesting to evolutionary psychologists because they can often be converted into reproductive success. Wealthier people in most preindustrial societies produced more surviving children. Wealth today has little relationship to reproductive success, however. In the past, wealth had more consequence for the reproductive success of men than women because it affected their ability to marry. This meant that when people made their wills, they might give more to sons than daughters because sons were a better vehicle for converting wealth into reproductive success. This was practiced particularly among affluent parents. Poor parents, whose bequests were too small to affect the marriageability of their sons, might prefer to invest in daughters who were more likely to have children than poor, unmarriageable, sons. The daughters could then invest these modest resources in their children.

The general argument that parents place more of their investment in the gender that is better able to produce grandchildren is known as the Trivers-Willard hypothesis. Reasonably well supported for some other species, this hypothesis has produced very mixed results for humans. One study of Canadian wills found that will makers behaved very much in line with Trivers-Willard. Rich stiffs gave more of their wealth to sons, whereas poor decedents preferred daughters. A study of Californian wills found no such pattern, however.[28] Californians treat sons and daughters equitably, regardless of their own level of affluence. The disparity in results may imply that legacies are unaffected by differences in reproductive potential between the sexes. Alternatively, it is possible that the prevalence of divorce and blended families today creates a great deal of diversion of resources to nonrelatives—stepchildren. Men are somewhat more likely to remarry following divorce so that there is greater potential for diversion of resources in the male line. We do not know whether such considerations affect will makers, and until such issues are systematically investigated, we cannot fully explain why the details of "testamentary behavior" (will making) are not consistent with some predictions drawn from kin-selection theory.

Although most will makers favor blood relatives, many also support their favorite charities, which tells us something about the human psyche. In varying degrees, most of us have altruistic tendencies in a pure sense that have nothing to do with favoring our own kin. We will next see that our more generalized altruism is the result of evolution in small groups composed of kin and nonkin that sank or swam depending on their ability to get along well together.

CHAPTER 2

Evolution and Ethics

If you want to study animal altruism, you should observe close relatives and expect altruism to flow from older individuals to younger ones who still have most of their reproductive careers ahead of them. Yet, altruism that favors relatives is not the only type that can evolve. It is possible for unrelated animals to help each other out, if there is some assurance that their good deeds will be paid back in future. You scratch my back and I'll scratch yours. This is known as reciprocal altruism. Of course, people may help complete strangers without any return for reasons that will become clear later.

Scientists are drawn to the study of reciprocal altruism because it is rational. If a benefit is returned at a future time, then there is really no net cost to the altruist, apart from delay and uncertainty of being paid back. Cooperation can also have immediate benefits. When lions cooperate to take down a large prey animal, they can all feed together, so there is no danger of anyone getting taken advantage of. Either way, animal cooperation can be a win-win situation. For this reason, biologists are often criticized for taking the altruism out of altruism. Yet, reciprocal altruism is of critical importance for our understanding of human morality. Its evolution plays a central role in grasping the evolution of human moral emotions such as guilt and outrage.

ALTRUISM IS NOT ALL IN THE FAMILY

When William Hamilton devised his gene-selectionist explanation for the evolution of kin-based altruism he pointed out that for altruism to "work" its benefits had to exceed the cost. The cost was measured by the number of copies of an individual's genes getting into the next generation.[1] In more technical terms, the benefits to the altruist's inclusive fitness had to exceed the costs. This idea is now referred to as Hamilton's inequality and is the origin of all kinds of rather silly scenarios such as: I should not be prepared to lose my life to save a cousin from drowning but if I can save nine cousins while losing my own life, my inclusive fitness gains and such altruism is adaptive.

These scenarios lack credibility because they almost never crop up in the real world. Yet, the conflict between benefits and costs of altruism is real. In general, helping distant kin, such as first cousins, is not a very good idea according to gene selection. Therefore, a meal that is diverted from a hungry son or daughter to a hungry cousin is not likely to contribute to an individual's inclusive fitness because the cost, in terms of lost direct fitness through the child (relatedness, $r = 1/2$), is four times the benefit through inclusive fitness via the cousin $r = 1/8$). Hamilton thus explains the fact that altruism between first-degree relatives is much more common than altruism among more distant kin.

Kin-selected altruism outside the immediate family ($r = 1/2$) is uncommon in the natural world. One possible example concerns the giving of alarm calls by Belding's ground squirrels at the approach of a predator.[2] This behavior helps other members of the social group to avoid being killed. It is altruistic because the alarm caller focuses attention on himself, which increases his own likelihood of being eaten.

Alarm calling by the ground squirrels appears to be determined by a fine gradation of costs and benefits. This is according to the degree of relatedness of altruist and beneficiaries and the loss of reproductive success of the altruist. Females are much more likely to give warning calls than males. This makes sense because females spend their lives in the same area and are related to many of the other ground squirrels. Males, on the other hand, leave their homes in search of mates, move into a new group, and are thus genetic outsiders. From the male perspective, the death of another squirrel will not affect his reproductive success, but preventing that death would involve some personal risk. It is hardly surprising that males almost never give warnings.

Among the females, older individuals are more likely to produce alarm calls than younger ones. This is interesting for two reasons. The first is that if an old female is never to reproduce again, risking her own life will not affect her inclusive fitness. The cost is close to zero so that even a small increase in survival and reproduction of kin will allow the behavior to evolve. The second is that very old females may have few, if any, surviving first-degree relatives so that they may be increasing their inclusive fitness purely through distant kin.

There is at least one other compelling example of nonhuman kin-selected altruism among relatives more distant than the immediate family. This occurs in the very odd context of cannibalism. Cannibals avoid eating relatives. If you were a tiger salamander and you could talk, you might say: "I saved all my brothers and sisters from a cannibal. I decided not to eat them."

Cannibalism is surprisingly common in the animal world. Salamanders are interesting in having two morphs, or types. The cannibal type is much bigger than average with large jaws that allow it to swallow prey almost its own size.[3] In addition to the broad snout, cannibals have hard bony plates holding long curved teeth that help them to grasp smaller members of their species. Researchers have found that the cannibal morph is less likely to develop when siblings are raised together.

Careful experiments show that cannibal tiger salamanders avoid eating siblings if unrelated prey are available. The cannibals can distinguish siblings from cousins and are more likely to feed on the cousins. They can even differentiate between cousins and nonrelatives, preferring to feed on the latter. This strange example may stretch the definition of altruism as commonly used. To a biologist, it fits the bill as an instance of kin-selected altruism for distant relatives (i.e., those less closely related than parents and offspring, or siblings).[4]

Apart from human beings, there are not many such compelling examples of kin-selected altruism outside of the immediate family. Such altruism may occur, but Hamilton's inequality (i.e., that the benefit in terms of copies of one's genes transmitted to the next generation must exceed the cost in the same terms) makes it rare. It is usually a much better idea to invest in one's own offspring than in more distant relatives, at least in terms of inclusive fitness benefits. Cooperative childcare among the Efe would appear to be a counterexample.[5]

Despite such apparently exceptional communities as the Efe, we expect reproductive-age women to invest most of their efforts in raising their own children, and very little in caring for the offspring of distant kin or nonrelatives, and that is generally true. Men do the same with one interesting twist that provides a fascinating example of kin-selected altruism at work.

Women have one distinct advantage over men in that they can identify their offspring with certainty. The proverb says that it is a wise son who knows his own father. It is also a wise father who knows his own son, at least in societies where married women customarily take lovers. About a quarter of women in the United States admit to at least one extramarital affair in their lives.[6] This makes adultery by American women rare compared with other societies. American men have high confidence of paternity and therefore invest heavily in children of their marriages. Or at least that is what happened in the recent past when most people were still marrying before raising children. Confidence of paternity is much lower in many of the societies studied by anthropologists, particularly those in South America and the South Sea Islands.

In such societies, it is common for men to withdraw much of their emotional and material investment in children of their marriages and focus their affection instead on nieces and nephews produced by their sisters. This phenomenon is known as the avunculate. Where this situation arises, the relationship between uncles and children bears analogy to the relationship between fathers and children in other societies.[7] The rationale in terms of gene selection is fairly clear. Where men father few of the children produced by their wives, they do more for their inclusive fitness by investing in their sister's children than in the children of their marriage. Rough math indicates that if a man fathers fewer than a half of his wife's children he is better off investing in the children of a full sister, from the point of view of inclusive fitness. In practice, if women have many partners, the sister is more likely to be a half-sister so that confidence of paternity would have to be lower than a quarter for married men for the avunculate to increase their inclusive fitness by investing in nieces and nephews.

Such extended-kin altruism is rare in the natural world. Altruism directed at strangers is rarer still, but the fact that it occurs at all is of great significance for understanding the evolution of human morality.

RECIPROCAL ALTRUISM: OH BATS!

Reciprocal altruism between unrelated animals seems unlikely.[8] To begin with, there is the question of trust. If you are nice to an unrelated individual today, how do you know that he will repay you, or reciprocate, when you want him to? The second issue is less obvious, but just as important. For reciprocation to work, it has to be balanced or fair. How is it possible for animals to keep a fair reckoning of what they have given and received? Even a cognitive skill for fair reckoning is not enough. Without some form of enforcement, reciprocal altruism is unlikely to work. Imagine a supermarket that operated under the slogan "Take all the supplies that you need today and pay us back whenever you feel like it." The monetary economy often requires immediate repayment to function smoothly.

If individuals are well known to each other, delayed reciprocation is possible. Thus, the food-sharing systems of many hunter-gatherer communities are based on delayed repayment.[9] This is a kind of insurance system in which the risks of hunting without success are spread across all the hunters in a group. Groups like the !Kung of the Kalahari Desert in South Africa hunt cooperatively. On most days, individual hunters fail to bring down large game animals. This is not such a threat to their families as it might otherwise be because they can draw on their insurance system. Successful hunters divide up their kills. Interestingly, this is usually done with complete fairness, as far as anthropologists can see. The hunter gives no more to his own family than he does to others. If kin were favored, the system would break down along lines of genetic relatedness.

The logic for sharing food among hunters is fairly compelling, at least in the case of large game animals. A man who is successful in the hunt has far more meat than he or his family could eat. Without any method of refrigeration or preservation, the meat would spoil in a couple of days and lose its value as food. Under reciprocal altruism, the hunter can invest this food excess into the reciprocal system and draw it out again over the many days when he is unsuccessful in the hunt but other individuals are.

Food sharing among hunters is a compelling example of a reciprocal system that benefits all of the participants. It is true that the very best hunters may contribute more meat than they draw out, but there are many days on which they fail in the hunt, so the insurance protection is vital for them too. Even good hunters may not bring down a large game animal more than once every couple of weeks. This would not be a viable method of feeding children. Good hunters thus come out better by participating in the system than if they hunted alone. The fact that they contribute more than they take out is not a complete loss either because their success at hunting increases their social status, which makes them more sexually attractive to women.[10]

Reciprocal altruism works well for stable human groups like hunter-gatherer bands whose members are well known to each other. Evolutionists have wondered whether this is due to human rationality or whether a similar system could be produced by natural selection acting on less intelligent animals. There is plenty of evidence that primates form cooperative alliances with other individuals who are not related to them. Young baboon males "adopt" juvenile females, for example. When the females mature, they breed with their protector.

Reciprocal altruism is pervasive in the lives of chimpanzees. These animals form alliances that play a significant role in their daily lives. Thus, two males can work together to intimidate a dominant male that neither would dare to take on alone. Chimpanzees also share food, although this is quite rare. Another service that chimps, and other primates, routinely perform for each other is grooming. They literally scratch each other's backs, removing skin parasites in the process. This is a low-cost activity that provides important benefits to the recipient and is a key feature of primate social life. In addition to its health effects, grooming can be socially beneficial because it helps individuals to relax and thus defuses possible aggression. Grooming is most often carried out between close relatives, or mates, but pairs of unrelated "friends" may groom each other too.[11]

In addition to grooming each other, unrelated chimpanzees have often been seen to embrace another who is hurt, or distressed, suggesting that they are capable of empathy. They seem to feel another's pain, so to speak. Animal behaviorists are often reluctant to draw such conclusions because they look unscientific, but experiments have shown that various social animals, including rats and pigeons as well as monkeys, will work to reduce distress in another member of their species.[12]

Despite such objective evidence, behavioral scientists are wary of attributing human thoughts and feelings to nonhuman animals. This scientific sin is referred to as anthropomorphism. In the chimpanzees' case, some anthropomorphism may be scientifically justified, given that chimpanzees are only about 1 percent different in genetic makeup from humans. After all, these animals are highly intelligent and learn to recognize each other in mirrors, according to careful scientific tests, something that is not possible for monkeys or the relatively small-brained gorilla.[13] (Orangutans also pass the mirror test.)

The prevalence of mutual helping among chimpanzees is fascinating in its own way, but it may tell us little about the problem of the evolution of reciprocal altruism. After all, it could be argued that chimpanzees, like humans, are smart enough to figure out the benefits of reciprocating. If so, an evolutionary explanation may not be required. Mutual aid societies could simply be a side effect of a high degree of intelligence, which is itself admittedly determined by brain evolution.

One of the most compelling examples of mutual helping comes from animals that seem far less intelligent than chimpanzees, vampire bats.[14] The bats live on blood that they steal from large animals such as cattle. This food supply is highly unpredictable, and individual bats may return to their caves without having fed at all. For adults, this happens about one night in fourteen, and for juveniles less than two years old about one night in three. Bats have a remarkably high metabolic rate which means that if they do not feed regularly, they will quickly lose body weight and die of starvation within a few days.

After a bat has fed, it can regurgitate blood to feed an individual that is starving. Some of this altruistic behavior may transpire between related individuals, but much is between unrelated bat friends. The logic is that food sharing is a kind of insurance system for bats, just as it is for human hunter-gatherers. Risk is better managed by distributing it outside the family.

The life of a vampire bat can be thought of as a race to avoid starvation. A fully fed individual has about sixty hours on the clock. One that has not fed for two days is in serious trouble, but if it is fed by another bat, as much as eighteen hours can be added to the clock. Interestingly, the bat doing the feeding loses only about six hours off its clock. Given that the benefits far exceed the costs, it is not hard to understand why bats would share food.

The only real problem is in understanding how an altruist gets repaid in its time of need. One answer is the establishment of stable friendships in which there is little chance of being rejected. Bat friends recognize each other by means of special contact calls. Even if both are well fed, they often get together and express their friendship through mutual grooming. Reciprocal altruism generally requires such stability and predictability.

Reciprocal altruism may stretch our understanding of animal behavior, not to mention psychology, but it is an attractive idea for scientists because it is pleasingly logical and balanced. By pooling their food stores, each bat contributes, and receives, about the same amount, and both parties in a bat friendship do much better than either would alone. But what about a situation in which an individual contributes much more than he gets back? Such unbalanced giving is rarely seen outside close relatives. Where it does occur, it might reflect an evolutionary mechanism promoting group survival, or group selection.

GIVING MORE THAN ONE GETS: GROUP SELECTION

The best way to grasp the idea of group selection is in terms of real examples. Group selection has been invoked to explain the emergence of multicellular organisms from single-celled ancestors; the evolution of nonvirulent strains such as for the rabbit disease myxomatosis; and the scarcity of males (who perform no useful work) in colonies of bees and wasps; as well as the reproductive altruism of termites, and of worker bees in colonies where the queen has several mates.[15] In each of these cases, it can be argued that the interests of the group take priority over the interests of the individual. In nonvirulent strains of myxomatosis, for example, individual viruses reproduce more slowly so that the host animal lives longer, permitting the entire virus colony living within a rabbit to survive. Such phenomena are a matter of ongoing controversy among evolutionary biologists, many of whom were trained to believe that group selection is a fallacy.

The best, and perhaps only, uncontroversial example of group-selected *behavior* concerns the cooperation of unrelated ant queens in founding a colony, a phenomenon that occurs for many unrelated ant species in a dozen genera. Such cooperation is adaptive because jointly founded colonies produce more workers and are thus more likely to survive in a competitive environment where as many as nine-tenths of colonies are wiped out in raids by other colonies. As soon as the workers emerge, however, queens fight to the death until only one remains.

Despite many claims, there is only one compelling example of group-selected *altruistic* behavior, and that could well change in the light of further knowledge. This concerns the foraging of a species of leaf-cutter ants, *Acromyrmex versicolor*, that inhabit the Sonoran desert of the US southwest and Mexico. Versicolor is the only ant species in which cofounding queens forage: the others seal themselves up until the workers emerge.[16] Moreover, field workers never observed aggression among the cofounding queens in this species.[17] These desert ants live in underground nests containing two or three unrelated cofounding queens and their offspring. Before the emergence of workers, the queens must go out themselves in search of food. Foraging is a dangerous activity that means leaving the relative safety of the nest and being exposed to predators above ground. In a particular nest, it is usually the same ant that does all the foraging. This looks like an individual forager contributing far more to group welfare than she receives in return. Biologists have long viewed such a phenomenon with extreme skepticism because it violates the expectation that individuals should be competitive, that is, that they should favor their individual interests over those of others in the group.

Yet, there is a very good Darwinian reason that an expert forager should continue to contribute to the other queens—and that they should care for her young when she is out foraging. All of the cofounding queens are likely to share a common fate. When Versicolor ants have raised their offspring, all emerge from the nest to engage in warfare against other nests in the vicinity. Far more ant nests are formed than the local ecology can support, so the groups that survive do so only by wiping out rival groups. As in the case of human warfare, the larger, better-fed armies are likely to prevail, all else being equal. This helps to explain why the forager specialist should put her life on the line every day for the benefit of the colony. Her investment in feeding the troops is essentially an investment in her own survival and that of her offspring.

Other explanations in terms of individual selection are possible, but they seem improbable. Thus, it is possible that Versicolor foragers are social subordinates who are being bullied and exploited by higher-status queens, but this interpretation is not supported by the evidence. Cofounding ant queens do not engage in a period of extensive fighting that would be necessary to set up a dominance hierarchy. Their interactions, moreover, are peaceful and evidently unaffected by a dominance system.

That this apparent example of group-selected altruistic behavior is the only good instance of its kind is revealing. Group selection is just not a very good explanation for altruistic behavior seen in nature. The reason is fairly simple: it is quite unusual for natural selection to operate principally at the level of the group. In general it is strong individuals who survive and pass their traits to offspring while their weaker competitors leave few descendants. In the Versicolor example, individual differences are far less important for reproductive success than differences in colony size and health.

Among human beings, there are a great many examples of extremely self-sacrificial behaviors from health-damaging overwork to military heroism, but a group-selection interpretation seems unlikely. As far as we can tell, our hunter-gatherer ancestors spent the last two million years in widely dispersed groups lacking the intensity of deadly group conflict that might have produced group-selected altruism. Surviving hunter-gatherer societies, like the !Kung of South Africa, have little or no warfare between local groups, although there is plenty of violence within groups.[18] Group composition changes frequently as well: if groups do not last for a generation, they cannot be subject to natural selection. In recent millennia, with rising world population, warfare has become more pervasive, thus creating a scenario in which group-selected altruism is a theoretical possibility with rival groups competing intensely over territory, just as the desert ants do. In terms of evolutionary time, this modern epoch is but the blink of an eye, probably too short for significant changes in our social behavior to have evolved.

Whatever the evolutionary debate over the sources of altruism, the mechanisms through which altruistic behavior is generated are even more obscure. We know close to nothing about the feelings of bats, much less ants, but there is every reason to believe that humans have special emotions, like guilt, embarrassment, and moral outrage, that evolved to ensure that we live up to our social obligations. The presence of such emotions in human beings has encouraged scientists from Charles Darwin[19] on to imagine that they must also be present to some degree in other creatures too.

MORAL SENTIMENTS

The concept of human beings as just another evolved species on this planet goes against the grain of previous human-centered ways of interpreting our experiences. It is exactly the same kind of deflation that people experienced when they discovered that Earth was not really the gravitational center of the universe. There are two distinct ways of dealing with this sense of letdown. Either we can focus on unique human qualities, such as language, or we can make the most of a bad situation by discovering wonderful qualities in other species that we had not previously suspected. Both strategies are fraught with problems. Language capacity is, after all, a biological adaptation, supported by specialized structures of the larynx and by evolutionary modification of brain structures that produce sounds, process incoming voice messages, and organize words into grammatical speech patterns. There is still some controversy about how all of this works, but work it does. Stroke victims often have speech problems following damage to their frontal lobes, for example, which demonstrates what can happen when even small language-specialized regions of the brain are put out of action.

If language is a human adaptation, it cannot be truly unique. Thus, evolutionary biologists see many similarities between human language and birdsong. Young songbirds may learn the local dialect of their song, just as children learn to speak with exactly the same intonation and stress pattern of the people they interact with. This is not to deny that human language differs in many significant ways, particularly with respect to its overall complexity, grammatical structures, and capacity to convey abstract meaning.

The alternative approach to human uniqueness is to discover the richness and complexity of animal life. This strategy was favored by Charles Darwin. Darwin's book *The Expression of Emotions in Man and the Animals* (1872)[20] explored the ways in which animals communicate emotions such as rage and fear. A dog that is frightened cowers, making itself seem as small and inoffensive as possible. When the dog is angry, its fur stands on edge, it snarls and bares its teeth, emphasizing its ability and willingness to do damage. Darwin was primarily interested in the usefulness of these displays for helping individuals to survive and reproduce. Fear is obviously useful for a dog that is outmatched by a rival because it can avoid a fight and possible death by showing it poses no threat. Similarly, a display of anger helps the stronger of two opponents to obtain a disputed resource without exposing itself to the grave risk of being wounded in an actual fight.

People express their emotions in somewhat different ways. Instead of raising our hair, we puff out our chests. Instead of baring our teeth, we clench our fists. Despite these superficial differences, there is no doubt that for people, as for dogs, anger may be expressed by positioning our bodies as though we were about to attack someone. Similarly, when a person is afraid, he expresses his fear by making himself seem small and inoffensive. The gaze is lowered, the shoulders are hunched, and the arms are held inoffensively close to the body.

The fact that animals have emotional displays that are intuitively so easy for us to understand suggests that they might have similar emotional experiences. Darwin could see no reason why not. After all, we share a long evolutionary history with other mammals and have brains that are fundamentally very similar to the general mammalian design, even if the human cortex is much larger relative to the rest of the brain.

The notion that there is continuity in the mental lives of humans and animals has had a checkered history. It has horrified academic psychologists. Their reaction can be summed up in the phrase "Don't go there!" Yet, comparative psychologists have been going there in droves, at least so far as investigating the minds of our closest primate relatives—chimpanzees, orangutans, and gorillas—is concerned. The whole impetus behind the laborious enterprise of teaching chimpanzees sign language, for example, has been an attempt to get in touch with the animal mind.

This enterprise was first taken up by George Romanes, who in 1883 published *Animal Intelligence.*[21] Romanes collected anecdotes about the intellectual feats of pets as told by their owners. The flaws of this method can be illustrated by an acquaintance of the author whose dog "talks" to her. This terrier makes observations such as "I love you," rendered as three barks.

Psychologists have spent over a century pointing to the scientific flaws of anecdotes. Imagining that our pet thinks about the world in the same way that we do can be a serious mistake. This scientific sin is known as *"anthropomorphism."* Students are taught that anthropomorphism, anecdotalism, and Romanes are very, very bad. Science may tell us that anthropomorphism is wrong, but it offers little explanation for the curious emotional closeness between people and their pets. In one survey, 48 percent of English respondents agreed that their dog was a family member, 73 percent let it sleep in the bedroom, and 40 percent celebrated its birthday.[22] Are the people doing all the work in these relationships, or is there a genuine meeting between the emotional life of the pet and the person? Decide for yourself.

Like Romanes, Darwin was a proponent of mental continuity, although he was well aware of the scarcity of relevant information. He imagined not only that animals must share some of the intelligence of humans, the theme developed by Romanes, but also that there is similarity in the emotional lives of various species. Darwin was even prepared to extend this argument to moral emotions like empathy and outrage.

He saw the ability to sympathize with another individual as a central "social instinct" of humans that could be found in all societies, however "savage," to use the term favored by the Victorian English for preindustrial societies. If such moral sensibilities had evolved in human beings, Darwin thought that it should also be possible to detect them in other species.

Like Romanes, Darwin was influenced by anecdotes about the apparent capacity of animals to feel for the suffering of other members of their species. He was moved by stories of blind crows being fed by other crows, for example. He took them at face value as possible evidence of empathy in other species. Such evidence was even more striking in the case of species that were closely related to humans in terms of their shared evolutionary history. Darwin was struck by a zookeeper's accounts of sympathetic acts by chimpanzees. If one animal was hurt, or upset, another would offer comfort by embracing it. Flimsy as this evidence might seem today, more careful work by contemporary researchers has drawn similar conclusions about the ability of chimpanzees to empathize with the sufferings of others.[23]

Such moral emotions may be necessary for altruistic behavior in primates and other mammals, as well as humans. Thus, the care of young by mammal mothers is probably based on intense emotions. Sheep mothers form a strong emotional bond with their lambs within a few hours after birth. A ewe becomes very distressed if separation occurs for any reason and will search frantically for the missing lamb, indicating that her actions are determined by emotions rather than a reflex response to stimuli that are present. Altruism directed at kin clearly has an emotional basis, although it would be a mistake to attribute any emotion to bees who take care of their sisters, if only because the nervous systems of bees are so much simpler. This simplicity restricts their actions to fairly basic stereotyped responses to specific stimuli.

Reciprocal altruism can be based on emotions as well. Evolutionary biologist Robert Trivers, a professor of anthropology and biology at Rutgers University in New Jersey, believes that the array of emotions experienced by people today is greatly affected by an evolutionary past in which we cooperated a great deal with individuals who were not close relatives.[24] The extent to which such emotions are found among other species is largely guesswork. If a bat has often fed a friend with regurgitated blood, and if the friend failed to reciprocate, we can imagine that the bat would be upset and angry, but we really have no idea what the bat is feeling.

Primatologist Franz de Waal describes an incident at his chimpanzee colony in Arnhem, Holland, in which a pair of individuals delayed the evening meal by staying outside after all the others had come in. When the "naughty" individuals were rounded up, the other members of the colony expressed their anger in such certain terms that they had to be housed separately for the night.[25] According to de Waal, the colony was still upset with the miscreants on the following day. Was this a kind of moral outrage, as de Waal believes? Or was the emotional outburst of the chimpanzees for some other reason? It is almost impossible to tell.

We do know that humans have a variety of moral emotions, such as guilt, embarrassment, outrage, empathy, and gratitude, that are expressed or otherwise manifested in all societies. Robert Trivers argues that some or all of these emotions evolved because they helped our ancestors to benefit from the practice of reciprocal altruism, such as that seen in the mutualistic food sharing of surviving hunter-gatherer societies.[26]

Each of these emotions has a role to play in our concept of fairness. Every society has a well-developed sense of justice, complete with some forum for airing disputes among individuals. In modern Western society this involves a legal system of endless, and mind-boggling, complexity. For hunter-gatherer societies, the system of justice may be no more than an argument between the aggrieved parties that is refereed by a headman. Whether the administration of justice is simple or complex, it is always based on a sense of fairness that has the same feel in different societies. If we help a person out in some way, we are normally entitled to expect some repayment and to feel aggrieved if the recipient refuses to return the favor. People who do not live up to their moral obligations excite anger and may be treated very harshly. In Western society, a man who is discovered stealing can be locked up for years. In some countries of the Middle East, he may have his hand chopped off. Evolutionists like Trivers believe that the emotion of moral outrage, and the aggression and punishment to which it can lead, are based on an evolutionary mechanism for persuading others to live up to their obligations. Other theorists believe that moralistic punishment as a cost to the punishing individual could not evolve through individual selection. It is therefore a case for group selection whereby punishing behavior evolves because it protects the interests of the group[27] (see chapter 7 for more on this controversy).

GUILT AND SHAME: MORALLY SHOOTING ONESELF IN THE FOOT

If hunter-gatherers share their meat equitably, then an individual who conceals a portion of his kill to ensure that his family gets more of the kill than others would seem to be doing himself and his family a good turn via improved nutrition—at least if he were not discovered and penalized. Yet, that is not how hunters behave, so far as anthropologists can tell. The worst that hunters do is to conceal meat from their wives by sending it surreptitiously to an extramarital sex partner. As far as the sharing relationship of the hunters is concerned, absolute fairness reigns. In a world of cynicism, it is always a shock to find that people routinely behave much better than they are supposed to!

Why are hunters so good at living up to their obligations to the group? Most likely they realize that to do otherwise would be to turn the whole hunting band against them. Moral outrage is a necessary enforcement system for reciprocal altruism among humans, and it plays an important role in keeping people honest. This helps us to understand why a capacity for moral anger should have evolved among our ancestors. People who did not respond with aggression when they were taken advantage of by partners in reciprocal exchanges would have been at a severe disadvantage. Others would have seen them as suckers and taken advantage of their weakness.

Moral outrage can be seen as the enforcement behind food sharing and other social exchanges. Other moral emotions are not so easy to explain. What about inwardly directed emotions like guilt and embarrassment?

The fascinating thing about guilt is that it seems so self-destructive. Why would people torture themselves over their perceived failure to live up to the moral expectations of others? Surely they would be better off looking out for their own interests.

Some insight on the social value of guilt is provided by the fact that not all people have the same capacity for feeling morally disappointed in themselves. Some individuals, particularly sociopaths, feel no sense of inner discord over their evil deeds and feel sorry only about being caught and punished for their crimes.[28] Also referred to as antisocial personalities, such individuals make up about half of the male prison population.

Moral explanations of crime have fallen out of favor, but when they were more popular sociopaths were referred to as *moral imbe-*

ciles. Some genuinely lack the capacity to feel empathy for other people. Most psychologists believe that a small number of individuals may be morally challenged for genetic reasons, but it is also true that children raised in deprived and abusive environments often have little opportunity to learn how to fit in with social expectations. If morality is thought of as a kind of intelligence, it cannot develop normally without the appropriate kinds of stimulation. Children must be rewarded for their good behavior and checked for their bad behavior if they are to learn the distinction.

Most hard-core career criminals would be diagnosed as antisocial personalities.[29] This gives us some idea of the problems that face individuals with a poorly developed sense of guilt. Of course, some of the brightest criminals never get caught. Moreover, many sociopaths are not criminals in a technical sense. Instead they inhabit the moral gray area of shady business. They are to be found in board rooms, systematically plundering the wealth of corporations. They wear the expensive suits of lawyers and politicians.

Despite the exceptions, the fact that sociopaths are to be found in prison yards tells us that even in the anonymous urban environments that most people inhabit today, being a moral imbecile is not likely to work out well. Sociopaths not only make everyone around them miserable, but also are vulnerable to leading unhappy lives themselves. It seems obvious that even in modern society persistent cheating simply would not work in an intimate community where everyone is very well known to everyone else. Hunter-gatherer societies certainly have their conflicts, and these conflicts are liable to end in deadly violence.[30] Yet, there is little anthropological evidence of free-loaders when it comes to food sharing.[31] Men who did not pull their weight were simply not tolerated in the hunting community.

Instead of trying to exploit others, the whole tone of life in hunting communities is set by the desire to contribute as much as possible to the group. Men who are poor hunters experience a great deal of social anxiety in relation to their hunting performance.[32] Being able to bring down large game animals has the same implications for social success among hunter-gatherers as bringing home a large paycheck has in our society.

If being incapable of guilt is a recipe for personal disaster within a small group, then the capacity to feel guilt must provide a social advantage that allowed it to evolve as a human attribute, because it is seen, to some degree, in the great majority of people in all societies. Robert Trivers, among others, believes that guilt is a useful, even necessary, quality in order for people to engage in reciprocal altruism.[33] Just as moral anger threatens cheaters with being excluded from an exchange system, so guilt is an internal device that prevents us from going too far out of line. If we did not feel guilt, then we would stray too far from what others expected of us. We would end up being ostracized, or worse. Our ability to survive and reproduce would be greatly reduced.

Among our ancestors in altruistic groups, the ability to feel guilt would thus have been a definite advantage. Guilt is unpleasant, but useful. It is rather like physical pain that is bothersome but prevents us from behaving in ways that damage our bodies. Similarly, a sense of guilt prevents us from doing anything that damages our reputation and social prospects.

So much for guilt! What about embarrassment? This is surely the oddest of all forms of emotional expression. Embarrassment can be defined as the social transmission of a sense of guilt.

Guilt may be useful in keeping us on track to meet our social obligations so that we are perceived as satisfactory partners in a mutualistic system. That inner voice telling us that we have not lived up to our obligations goads us into doing better in the future so that we do not offend our allies. Guilt is useful, but what possible advantage can be gained from telling other people that we have not lived up to their expectations of us, which is what embarrassment does?

A familiar example of embarrassment occurs when a person who is not skilled at lying is forced to do so by social circumstances. A woman is invited to go to the movies by a female acquaintance whom she has known for a long time but secretly dislikes. Now she must come up with a plausible excuse that lets the other person down gently. The glimmerings of a possible prior engagement come to her quickly. She believes that is the evening she must help out at the church fund-raiser. Surprised at the fluency of her own response, and even a little triumphant that she has come through with a workable excuse, she is a little peeved that the friend seems unconvinced. Why? Well, her voice quavered as she stammered out the prior engagement. What is more, her normally pale face flushed bright red. Just as a flashing red light alerts motorists to danger, a face that suddenly goes red transmits the signal "I am deeply embarrassed by what I just said. It is probably a complete lie and you would do well to disregard it."

Why on earth would we communicate such wildly conflicting messages through different channels? The simple answer is that maintenance of the relationship may require face-saving lies. These are called white lies because they are motivated by a good intention: to avoid hurting another person's feelings unnecessarily. Some people, particularly kindly people, may find such lies difficult to pull off. Polished liars can deliver them with aplomb. The real question is why are some people such poor liars?

Communication experts refer to blushing and stammering as "nonverbal leakage."[34] In this case, the signal being transmitted through nonverbal channels is so strong that it completely overwhelms the content of what the woman says. There is no mystery about why the woman is embarrassed. She is embarrassed because she is lying. The real mystery is an evolutionary one. Why would we come equipped with a sort of alarm that occasionally goes off, saying, in effect, "Hey, everyone, I'm lying now! You can take everything I'm saying with a large grain of salt"?

Economist Robert Frank of Cornell University believes that the evolution of the human display of embarrassment was specifically designed to make people more appealing as social partners in an exchange system.[35] The logic is fairly elementary. Any kind of social exchange involves trust. If individual B benefits from the goods contributed by A today, how can A be sure that B will reciprocate when her turn comes around? On the surface, this is an unsolvable problem. All reciprocal systems have the potential for cheating.

Money is a perfect example. When we give someone a bank note in return for goods or services, this is a guarantee that he can receive goods and services of equivalent value in the future. Money promotes social exchange, known as commerce, because minimal faith is involved. We do not have to trust, or even know, a person to sell them something. Yet, even monetary transactions are open to cheating because a small proportion of the notes in circulation are counterfeit. Yet, if the cheating is kept to some low level, the monetary system works perfectly well because the cost of cheating is spread thinly throughout the economy. Moreover, business owners who fear receiving large numbers of worthless counterfeit notes that will be rejected by the bank can protect themselves by carefully checking all of the notes they take in, a laborious process that deters cheats.

Even if a reciprocal altruism system can tolerate some level of cheating, it works much better if there is some method of excluding cheats. Robert Frank believes that embarrassment evolved in humans precisely to provide a guarantee of honesty.[36] The logic is that emotionally transparent individuals are more attractive as exchange partners than people who can lie skillfully without any embarrassment. If people become embarrassed when they lie, their lies don't "count." We do not have to worry about them cheating us because they are emotionally incapable of doing so without tripping the alarm bell of embarrassment.

People who become embarrassed easily are attractive as exchange partners because they seem honest and trustworthy. Given the importance of cooperation in the survival of our hunter-gatherer ancestors, a capacity for embarrassment would have been a significant social asset, Frank believes. It would have served as a badge of honesty that gave individuals transmitting this message an advantage over those who were less emotionally transparent, and therefore a greater risk for reciprocal exchange.

Frank sees honesty and concern with the reputation for being trustworthy partners as critical for our hunter-gatherer ancestors. This argument is certainly consistent with what we know about the scrupulous fairness of surviving hunter-gatherer societies. Are the rules that we apply in dealing honestly with those we know well and deal with every day the same as those governing the interactions of complete strangers? You might expect that strangers would be treated less kindly than friends, or familiar others. According to Frank, however, there is a surprising degree of continuity between our economic interactions with complete strangers and those with acquaintances. One of Frank's favorite examples is the practice of tipping waiters in distant cities that are visited only once in a lifetime.

According to a rational economic analysis of tipping, people tip waiters following a meal because they expect the waiter to provide them with good service in the future. Clearly, this analysis is invalidated in the specific case of one-time visits to distant places. If no future service is anticipated, why do people provide a tip? Frank believes that they do so because of an evolved concern to protect their good name and reputation.

When someone sits down at a table in a restaurant, she enters into an implicit agreement to pay the waiter for services rendered. Failing to do so is damaging to her good name. It means that she cannot be relied upon to hold up her end in a bargain. This is relatively insignificant for a modern traveler moving anonymously between cities, but it would have been disastrous for our ancestors who lived in small cooperative groups where everyone knew everyone else very well. Even a single minor act of dishonesty in a social obligation could have been enough to shatter the cheater's reputation. Hence, according to Frank, we've evolved the punctilious concern with reputation that compels people to live up to an obligation in the modern world, even when there is no negative consequence for cheating.

Frank's explanation rings true, but it is incomplete. Surely, people tip waiters because of a sense of fairness: they know that wait staff are badly paid and rely on tips to make a living. Someone who does not tip the waiter in a society where this is customary is effectively robbing him of his wages. A sense of fairness may be as significant as people's concern about their reputation, but this distinction may be rather superficial. In a society where mutual exchange is based on trust, individuals need to be mindful of what others feel about some transaction, as well as how they are perceived as a consequence. In other words, they need to treat people fairly if they are to maintain reputation and trust. Tipping is an example of everyday reciprocation of money for services. It is vulnerable to cheating and can thus be considered an example of reciprocal altruism when it occurs. The gift of an organ, by contrast, is a once-in-a-lifetime instance of heroic giving that is never reciprocated. Why would people ever donate a vital organ?

BUDDY, CAN YOU SPARE A . . . KIDNEY?

Students of organ donation have studied the motivations of the donor.[37] Many organs are donated to organ banks rather than to individuals. Organs are typically harvested at the point of death so that their loss does not harm the donor in terms of lost health or quality of living. From that perspective, organ donation can be interpreted as part of a mutualistic system in which people who pay into a communal supply of organs are repaying the system for any organs that they, or their close relatives or friends, may need while still alive. This is analogous to the hunter receiving food from the communal supply today in expectation of repayment in the future.

Using that analogy, donating a vital organ after death is no different in principle from giving blood. Donating blood is another mutualistic system that is maintained by the generosity of individuals. Blood banks often complain about the dwindling supply of blood donations. This is a side effect of the general decline in volunteerism in recent decades, as the pace of life has quickened and people find themselves with less spare time. Despite this, the system has remained viable so that donations meet demand. Admittedly, the blood that is used may contain artificial components, thus stretching the supply, and some people may be paid for their blood donations. Yet, the blood donation system is largely voluntary and self-sufficient.

Eliciting blood donation continues to be a challenge[38] because there is no connection between a person's paying into the system and his ability to profit from it when he has a medical emergency that uses donated blood. The system, in a sense, is wide open for cheating. Most people do not cheat on waiters in foreign cities because they have built up a pattern of behavior involving payment of tips that controls their actions in different contexts. Conversely, most people do not have a pattern of regular blood donation. Giving blood can be an inconvenience, takes a little time, can be somewhat painful, and requires the completion of a humiliating questionnaire that delves into a person's sexual activities and use of illegal drugs. In short, the costs of contributing to the system often seem greater than the costs of cheating. Consequently, the great majority of people "cheat."

Organ donation by living people constitutes a rare, but ethically fascinating, example of giving that is necessarily out of proportion to any possible benefit received. In the animal world, such heroic sacrifices are normally made between kin, usually from mothers to offspring. The fact that most living organ donations are between family members is hardly surprising from this point of view. Yet, there is less here than meets the eye because the organ has to be immunologically compatible with the recipient. Surgeons also prefer to use organs from young, healthy individuals when possible, which makes it more likely that children will donate to parents rather than vice versa, thus going against the usual path of investment from parents to offspring.

The really fascinating ethical conundrum arises when a sick person needs a compatible organ, such as a kidney, from a close relative to survive. Kidneys come in pairs but only one is necessary for healthy survival. The downside of donating a kidney is that if the remaining kidney fails, there is no backup and the donor may sicken and die if he does not receive a donated organ himself. The question is, does a person with a large contingent of young, healthy relatives have an advantage in his hour of need for a donated kidney? Social psychologists who investigated this question were astonished to find that a person who had only one relative qualifying as a potential organ donor had a much better chance of receiving the donated organ than someone with many healthy close relatives.[39]

This conclusion defies mathematical probability, but it is fairly easy to explain in terms of common sense. If you are the only possible donor, all of the responsibility descends on your head. You and only you can save the life of the sick relative. You can do so only at some cost to yourself, but this cost is much less than the benefit to the recipient. In effect, refusing to donate is ethically equivalent to letting the sick person die.

When there is a large pool of possible donors, responsibility is spread out among them. Social psychologists refer to this as diffusion of responsibility. It is the reason that if someone is mugged on a crowded subway train he may receive no help from bystanders. Everyone feels that someone should be doing something, but no one feels it should be him or her.

In the worst-case scenario, when no healthy relative has stepped up to provide the kidney and the patient dies, there is no single individual who must accept the blame. All may feel bad, but they can always think of compelling reasons why someone else should have been the donor: She was healthier. He had no children. The sick person had done more for her.

As altruistic as people are capable of being, their self-sacrifice is usually limited by the context. In the case of kidney donation to a relative, the conflicting needs of the potential donors may prevail over the interests of the person in need of an organ. This reminds us that people, like other animals, were designed by natural selection to protect their own interests in the face of the conflicting interests of others. In this case, the potential donors were saying, in effect, why should I make this sacrifice when there are many others who would be more appropriate donors?

When the chips are down, no one wants to be the loser. In other words, competing interests undermine altruism. Just as few people are willing to give up a kidney, we might expect them to be unwilling to forgo sexual behavior and reproduction in the manner of worker bees. What about those members of the clergy who sacrifice their Darwinian fitness and give up reproduction entirely by voluntarily abstaining from sexual intercourse?

CHAPTER 3

Sterile Castes of Priests and Nuns

Why was this furor of voluptuousness inflamed in our veins which arouses our blood and tortures us even more than it elates us, and which, under the deceptive name of love multiplies crimes among men?

Charles Perraud, from *Meditations on the Seven Words of Our Lord on the Cross* (1890)

Biologists assume that people, in common with other evolved life forms, are selfish. That is, they compete for food, shelter, and other resources that help them to survive. In addition, they compete for opportunities to reproduce. Nonhuman exceptions exist, however. Significant examples included food sharing among bats and other species, and the phenomenon of sterile worker castes among the social insects.

Such exceptions seemed like an insurmountable problem for evolutionists. Darwin felt himself defeated by the social insects, for example, and some scholars believe that his doubts on this score were a major reason that he sat on his idea of evolution by natural selection for some twenty years between his return from the *Beagle* voyage and publication, in 1859, of *On the Origin of Species*. Like everyone else, scientists love puzzles. In fact, attraction to the piece of evidence that doesn't fit often compels them to conduct further research, or change their theories, or both.

Conspicuous examples of altruism among other species have forced scientists to come up with elaborations of the nature-is-red-in-tooth-and-claw version of natural selection. Altruism in the sterile honeybee is no different from altruism of a parent toward offspring. Both are examples of genes for altruistic behavior transmitting copies of themselves to future generations. The only difference is that genes for parental altruism are transmitted through direct reproduction, whereas genes promoting sterility in workers are passed on through the queen, or via indirect reproduction. As noted, other kinds of altruism have been demonstrated in terms of animal examples, including reciprocity, or mutualism. There are a limited number of good examples of animal reciprocity, like vampire bats, but countless human examples exist.

It is a strange, and even shocking, irony that many of the atrocities committed by one human group against another can be explained as the obverse of reciprocal altruism within groups. Thus, soldiers make a personal sacrifice to promote their country's interest or that of their political or religious faction. At an extreme, young men may give their lives in a military campaign that has little or no chance of success, such as the Charge of the Light Brigade memorialized in Alfred Tennyson's jingoistic verse.

This brings us to the caste of celibates in religious orders and communities around the world that stretch back through recorded history. Priests, monks, and nuns generally obey at least one of the basic rules of life: they survive. Many renounce sexuality and reproduction, however. In some ways, this might seem less biologically anomalous than suicidal military missions. From a psychological perspective it may be even more so. A person can commit suicide only once, although soldiers may embark on potentially suicidal missions almost every day of their service in a war. Celibate priests must resist sexual temptations every day of their adult lives.

The rationale for priestly celibacy is less clear-cut than one might imagine. Most Catholic theologians believe that renunciation of sexuality helps priests to focus on their beneficial mission to parishioners, or that it helps monks and nuns to achieve a state of grace that is theoretically beneficial to the wider religious community. If these claims are correct, then religious celibacy is an interesting combination of bees and bats. The failure to reproduce evokes the sterile worker castes of bees. Putting the spiritual needs of others before one's own emotional needs is analogous to bats feeding their famished companions. Can kin selection or reciprocity shed any light on religious celibacy?

Anthropologist Hector Qirko of the University of Tennessee conducted a survey of celibate groups around the world.[1] Some type of socially recognized celibacy occurs in approximately nine-tenths of the preindustrial societies described in the Human Relations Area Files. In about one-fifth of these societies, celibacy occurs in an individual context with little connection to religious institutions. One example, among the Ingalik of the Yukon, is the *eniglani*—individuals who renounce marriage and withdraw from their communities. Another is the *karongs* of Palau (prior to European colonization), who claimed direct contact with the gods.

There are many examples in which celibacy provides benefits to kin groups if the celibate individual contributes to the reproductive success of other family members, often by contributing to food production. Examples include the Irish, Basques, and Lapps, where the burden of celibacy usually fell on later-born sons, although daughters sometimes delayed marriage to care for younger siblings. Such cases of celibacy may be easily interpreted as examples of kin-selected altruism, although it is easy to imagine the sorts of manipulation and competition among family members to which this might lead, as reproductive members of a family sought to delay or prevent the marriage of their "helper at the nest."

Close to two-thirds of examples of celibacy around the world occur in settings where celibates congregate with nonrelatives, mostly in Christian or Buddhist institutions. Other examples of permanent abstinence include: the priestesses of ancient Babylon (around 1750 BCE); the Vestal Virgins and the Cynic philosophers of Rome (both around 100 CE); the Pawnee Children of Iruska; the Inca Virgins of the Sun; the Aztec priestly class; and the nineteenth-century "Amazons" (i.e., female warriors) of Dahomey.

When religious institutions foster celibacy, they often do so in part by forming quasi-familial relationships and referring to each other using kin terms (*brother, sister, father,* etc., see chap. 8). This process may be facilitated by early initiation, and in many societies children are inducted into religious orders before reaching maturity. It would not be possible to discuss the reproductive altruism of all religious celibates in detail here, but the fitness issues can be illustrated in terms of an example that is somewhat familiar to most readers, namely, Catholic religious celibacy.

A BRIEF HISTORY OF CATHOLIC CELIBACY

Celibacy had a modest beginning in the early Christian church.[2] When the apostles were called to their ministry, many were married and famously abandoned their homes, wives, and families to follow the Messiah. During their later ministry, following the death of Jesus, however, some probably traveled with their wives and families. The early church was rather ambiguous about sexuality and had rather little to say on the subject. Saint Paul considered celibacy more blessed than marriage, and was celibate himself, but recognized that sexual desires can be a distraction for religious people that is lessened by marriage and sexual gratification. He concluded that it is better to marry than to burn in the fires of lust.

Celibacy was not expected of church leaders in the days of the apostles, but some adopted it voluntarily. Many of these had already been married and had already produced children. (The same pattern of adopting a religious vocation late in life after a family has been produced is seen in other contemporary religious traditions. It is particularly common in India, for example.) The fact that children had already been produced makes this pattern of celibacy less "unnatural," that is, less challenging to an evolutionary approach to human behavior.

There is little sense of a rejection of sexuality in the early Christian church. Women were not feared or seen as evil. Being married did not disqualify the apostles from religious service. The mere fact that Jesus liked to hang out with Mary Magdalene, who may or may not have been a prostitute, is a rather charming indication that the early church was at peace with human sexual appetites. A general lack of austerity is indicated by fondness for feasting and wine, which is sharply at odds with the practices of true ascetics, such as John the Baptist, who lived in the desert, where he was not bothered by publicans or sinners and subsisted on grasshoppers and wild honey.

Strong Christian rejection of sexuality first surfaced in the second century. Beginning with Ignatius (born in 100), several prominent church figures renounced sexuality and advocated celibacy for priests. They included Justin Martyr (born 150), Athennagorus (born 180), and Origen (born 185). Origen is notorious for having been voluntarily castrated to quell the fires of lust. Few men have been willing to go so far for their beliefs. That is why people were so horrified to discover that members of a contemporary doomsday cult, Heaven's Gate, had followed Origen's example.

A long line of ecclesiastical killjoys ensued. Perhaps the most influential of these was Augustine (born 344). Augustine propounded the astonishing view that all forms of sexual expression are bad and polluting. Even sexual intercourse in marriage is wrong if it is indulged in for pleasure. He argued that the purpose of sexuality was reproduction and that its use for any other purpose was contrary to God's plan. Augustine's views were echoed by other church leaders, including Jerome (born 340) and Gregory the Great (born 540). If sexuality was polluting for ordinary people, who lived decent lives, then it could not be countenanced for priests, who performed sacred rites, such as handling the communion host, that called for ritual purity.

The exceptional austerity of Augustine's views are difficult to explain due to the incompleteness of historical information on such matters, but we know, at least, that he was not alone. Many fourth-century religious men abandoned the comfortable life of cities to rediscover a simpler life in the wilderness where they could focus on prayer and meditation. The monastic tradition encouraged celibacy as a way of preserving singleness of purpose. It was necessary to renounce all earthly pleasures, including sex, to be completely focused on God.

Despite the rationale for why religious people should be celibate, celibacy was not a requirement for priests in the early church.

In fact, the theological controversy about priestly celibacy continued for many centuries without resolution. Some priests were celibate but others were married.

Married priesthood came to a sudden end in 1139 when the Second Lateran Council issued a decree that nullified all existing marriages of priests and declared that only single men could be ordained. This meant that a man's marriage had to be nullified before he could enter the priesthood.

The fascinating thing about the Second Lateran Council decree is that it had nothing to do with hatred of polluting sexuality, with unity of religious purpose, or with exclusive devotion to pastoral care, which are the primary theological arguments for celibacy. It was entirely a matter of economics. The problem was that married priests bequeathed their property to their children. In the process, parishes were being brought to the verge of bankruptcy. The decree was designed to choke off this financial hemorrhage.

The Roman Catholic Church reaffirmed the importance of celibacy at the Second Vatican Council (1962–1965) and again at the Synod of Bishops in 1971 in Rome. The rationale is that celibacy helps priests to concentrate on their religious vocation and on the needs of their parishioners. Despite the unwavering position of the Catholic leadership, support for a celibate priesthood has fallen among the laity. There is a widespread view that if people voluntarily opt for celibacy, they must be sexually dysfunctional. This view has hardly been helped by the current outbreak of scandals concerning priestly pedophilia (see chapter 9).

The widely held belief that celibates are sexually abnormal has hurt recruitment to the priesthood.[3] Young men are reluctant to admit that they want to be priests because their peers consider this an unacceptably deviant career path. Families no longer encourage their sons to enter the priesthood, something that was common in the past. Having a priest in the family was a sign of respectability and may have helped family members to confront difficult life changes, including bereavement. Disillusionment with priests, and with the priesthood as a career, has contributed to a serious and growing shortage of Catholic priests. Many Catholic laypeople recognize that the problem can be solved only by ordaining married men and admitting women to the priesthood, a transition that has been made by most of the world's major religions, with the exception of Islam. The church hierarchy is unwilling to accept these breaches with tradition, but it would appear to be holding out against the inevitable.

Celibacy is by no means peculiar to the Catholic Church or Christian traditions. According to Elizabeth Abbot in *A History of Celibacy* (1999), celibacy is, or was, practiced for religious reasons among Buddhists, Hindus, Jains, Native American shamans, Incas, and Naditus of Babylonia (cloistered women dedicated to the sun god Samas).[4]

SHAKER CELIBACY

One of the most extreme examples of rejection of sexuality was the Shaker sect founded by Mother Ann Lee, who was born in Manchester, England, in 1736, under conditions of the bleakest poverty and overcrowding. Her aversion to human sexuality began early as she shared a bed with her parents and witnessed their sexual intercourse. During the day, she pleaded with her mother to refuse the father's sexual advances. Ann was subsequently forced to marry and bore four children, all of whom died. Given this dismal personal history, it is perhaps understandable that she came to see sexuality as the root of human suffering and degradation.[5]

Having renounced marital sexuality over the reasoned theological objections of her husband, Abraham Stanley, a blacksmith, and the clerics he recruited to help him, Mother Ann experienced a vision in which she beheld Adam and Eve copulating in the Garden of Eden before being expelled by God for this act. From this vision, she reworked the biblical myth so that sexuality, rather than eating from the Tree of Knowledge, was the cause of humankind's fall from grace.

Mother Ann's antisexuality message had interesting feminist implications. Many of her female contemporaries feared pregnancy for good reason. Producing children was dangerous and difficult, and married women had an extraordinarily onerous life of continuous child production and childcare interspersed with illnesses and bereavements. Escaping sexuality was a way of escaping their lowly status as women. Celibacy was less popular among the secular and religious authorities in England, so the sect was persecuted. It ultimately moved to the New World, establishing its New Lebanon colony near Albany, New York, in 1774.

A colony of men and women devoting their lives to the cause of celibacy is surely one of the oddest phenomena of human history. Most religious traditions have recognized that when healthy young men and healthy young women live together, they form sexual relationships and will inevitably have sex. That is why people in most monastic traditions accepted sexual segregation. Even then, same-sex relationships tended to be eroticized, a problem that was acute in male monasteries. Deprived of female companionship, monks often invested their sexual feeling in each other. The medieval monk Pachomius, who established a monastery at Tabennisi, in Egypt, ate only once every three days to quell the fires of lust. He established a system of rules to maintain chastity. Above all, he was preoccupied with preventing the monks from tempting each other. The regulations were bizarre, almost Monty Python–like in their details. Monks should be careful to cover their knees when they sat down together. They had to be careful not to hike their tunics too high as they bent down to do the laundry. They had to avoid looking in the face of another monk. Never visit another monk's cell. Never lend a book to another. Never help him by pulling a thorn from his foot. Never hold hands. Never bathe with another. Never help him to oil his skin.[6]

The Shakers devised their own ingenious communal methods of sublimating sexual passion. They had meetings in which people publicly confessed their struggles against passion. These were steamy and specific and have been described as a verbal peep show. Another important method for exorcising lust was the frenzied dances in which the Shakers engaged (Shaking Quakers, hence "Shakers"), thus working off their sexual energies in the company of the other sex (see fig. 3). Shaker women also dressed in dull and dreary clothes that presumably prevented them from exciting lust.

Fig. 3. The Shakers, a religion that prohibits sex and therefore relies on proselytism for its survival. (Library of Congress)

The Shakers made many new converts in their early period in the United States, but the self-denial movement that arguably improved the lives of poor participants early on was running contrary to the historical current of increasing prosperity: as economic opportunities improved in the outside world, life inside the sect became progressively less appealing to potential recruits. By 1826, the sect had expanded to six thousand people living in eighteen communities. As a celibate society, they relied on proselytism to keep up their numbers. With the decline of religious revivalism after about 1840, the number of new converts fell steadily. What is more, children of converts mainly chose to leave the austere communities which were not only sexually egalitarian but denied that there was any special relationship between parents and children. This virtually guaranteed that young people would feel alienated and rebellious.

Moreover, the Shaker communities became quite prosperous thanks to their hard work and creativity. Their increased affluence made family life more desirable to some members because the risks of pregnancy and childbirth that had made such an impact on the founders are risks that are greatly worsened by poverty. During their religious ceremonies when members became possessed with the spirits of past leaders, they often spoke out against celibacy. Clearly, the movement had begun to fall apart from the center. With the drying up of the stream of converts and the increasing desertion and deaths of their aging members, the religion practically ceased to exist by 1980 and lives on, in name only, in the form of a few aged survivors living in Maine.

The Shaker movement is a fascinating social experiment from many different perspectives, but it is an instructive example of the very obvious principle that celibate communities cannot survive without continual recruitment from outside. Monasteries and convents have always had the same problem, of course. The Shaker community was a monastery that housed both sexes simultaneously while maintaining sexual chastity. That is a remarkable, even if misguided, achievement that may not have been accomplished in the previous history of the world. People were designed by natural selection to reproduce and not to live in celibate communities.

If this is true, how can so many religious groups defy this most important of biological imperatives by voluntarily refusing to have sex, assuming that the majority have normal sexual desires? The easiest explanation, but not necessarily the most correct one, is that priests, monks, and nuns are not really celibate, that they only pretend to be.

IS CELIBACY A SHAM?

The Catholic Church advocated celibacy among priests and monks for many centuries before formally banning marriage for priests in the twelfth century, as already noted. The question is, were priests really celibate, or was celibacy part of a holier-than-thou pose to be worn in public like the priestly vestments? Historical research indicates that many priests of the medieval period were far from chaste. Some of the evidence for this is compelling, including the existence of acknowledged offspring, but much is indirect and therefore not

The presence of such deviants hardly invalidates the sincere celibacy of the majority of priests. Peter de Kalbs was a bad apple; still, the entire barrel may have been quite corrupt. Thus, a quarter of the monks (11 of 42) had children and half were reported to be sexually active (20 of 42), according to depositions at the inquiry.[13]

Peter de Kalbs may have been something of a sex maniac, but he is far from the only one who took religious orders. Another Don Juan was the priest Pierre Clergue of Montaillou, a swashbuckling adventurer of the early fourteenth century who belonged to a powerful and ruthless family. Pierre devoted his life to hunting, spying, drinking, and fornication. Records of his inquisition, carried out by Jacques Fournier, the bishop of Pamiers, describe twelve mistresses. Clergue was accused of deflowering a fifteen-year-old virgin in a barn and then marrying her off to another man who tolerated the priest's sexual advances to his wife for four years. At the inquisition, this mistress claimed that she had been afraid to end their affair because she believed that the priest and his brothers were capable of murdering her, or otherwise obtaining revenge. Rotten as some of these medieval priests and monks may have been, it is very difficult to judge whether lecherous men of the cloth were any thicker on the ground then than pedophile priests are today.

In the days before reliable contraception, sexual relationships almost inevitably resulted in the birth of children. In one way or another, the church, that is, the parishioners, ended up paying for their support. If priests fathered children by their housekeepers, these children were often supported directly from church funds, whether they lived in the priest's house or not. Clerics may also have used their salaries to buy land for their illegitimate children. Instead of altruistically refraining from reproduction in the service of the parishioners, some priests did exactly the opposite. They foisted off the material support of their illegitimate children onto parishioners.[14]

Perhaps the most interesting, and conspicuous, example of rogue priests milking the church for all it was worth was the practice of getting their bastard children admitted to holy orders. This required a dispensation, which was effectively given by bishops acting in the name of the pope. This practice was widespread: In the archdiocese of Cologne, some sixty dispensations were given to priests' sons between 1310 and 1352, allowing them to be ordained. In the twelfth century Pope Paschalis II observed that most of the priests in Spain were married and that the majority of priests in England were themselves the sons of priests. Interestingly, he observed that the progeny of priests were temperamentally more suited to being priests.

Despite many official warnings against sensuality and the removal of a few priests from office for their sexual activities, it is clear that many priests and monks in the early church were sexually active and produced children. Although they were forbidden from having unrelated women stay with them, many clerics not only had concubines but even maintained many women simultaneously. These practices changed following the Second Lateran Council, but they changed very slowly. Thus, Henry of Gelders, a fourteenth-century bishop of Liege, bragged about having sired fourteen sons in less than two years.[15] Even allowing for some error of exaggeration, it is likely that he was sexually active with a very large number of women. Presumably he used the power and wealth of his office to make such phenomenal reproductive success possible. Instead of altruistically refraining from reproduction to minister to his flock, it is likely that Henry of Gelders did the opposite. As others did, he likely passed on to parishioners the responsibility and cost of raising his illegitimate offspring. Instead of being a reproductive altruist, in the sense of sterile worker bees, he played the role of a reproductive parasite, exploiting the power of his office to further his individual reproductive interests.

Reading about the sexual exploits of medieval priests and monks is rather shocking to modern sensibilities. We expect priests to live up to their professional ideal of chastity, even though we might have trouble doing the same ourselves if placed in that situation. Very few modern priests are married or have stable cohabiting relationships. As far as we can tell, most Catholic priests refrain from sexual relationships with women for much, or all, of their lives. That is why even a few exceptional cases excite such interest. One of the most infamous in the post–Industrial Revolution world was that of French priest Charles Perraud, who maintained a secret relationship with a childless widow for fifteen years until they were separated by her death in 1887. Perraud died in 1892, but his unusual life was not revealed to the public until 1908 in a biography by Albert Houtin titled *A Married Priest* (*Un Pretre Marie*).[16] It was the second biography of Perraud, who was seen as a model priest and who had made a name for himself as a devotional writer. Houtin's biography was sympathetic and it described his devout life and accomplishments as a priest. In addition, it included the struggle between wanting to be a good priest and wanting to be a married man and described his secret life shared with Madame Duval in their apartment.

The Perraud case generated a great deal of scandal because it suggested that the public image of well-known people is often a façade of lies and deceit. It created quite a buzz in the gossipy salons of the wealthy and set off a feeding frenzy in the popular press. Apart from the lies, there was the human-interest angle of an otherwise conscientious priest forced into a deceitful existence because of the unreasonable demands of celibacy.

Religious people are human beings and feel the same needs for physical and emotional intimacy as other people. Yet how are those needs appropriately managed or satisfied? That is something on which priests, nuns, and monks receive minimal guidance even today, much less in earlier centuries.

Most of the religious violation of the code of celibacy involved priests and monks. Nuns have generated far fewer scandals, perhaps because women in general inhibit their sexual impulses. One exception is the scandalous relationships between priests and nuns in seventeenth-century Venice. According to the Council of Trent (1545–1563), nuns were required to break all ties with the outside world. Venetian nuns got themselves into trouble by conducting long-term love affairs with priests.[17] The relationships were monogamous and intense but rarely involved actual intercourse. If even these Platonic attachments among celibates were outlawed, one wonders what priests and nuns were expected to do with their evolved emotional needs.

There is an interesting connection between the power of the church and apparent changes in adoption of celibacy by the clergy. In medieval times, the church was very powerful, not just in its hold over the spiritual lives of ordinary people but also in its access to their property. The church was a great owner of parish property and its coffers were constantly being replenished by obligatory church taxes, known as tithes. High-ranking clergy were wealthy and lived in great style. They belonged to the elite, which produced the most religious scandals. Wealth and celibacy have never mixed well not only because rich men are capable of purchasing sexual favors, but also because their power makes them more sexually attractive to women. Women reportedly brought their daughters to see Peter de Kalbs so that they could discover "the recipe of his sexual magic." It is hard to determine what exactly this meant, but it seems that women were attracted by his sexual vigor, just as females in other species are attracted to alpha males that sire a disproportionate number of offspring. Wealthy flamboyant church leaders seem to have been babe magnets! That is no surprise to evolutionary social scientists like Laura Betzig, who find that high-status men everywhere have an advantage in attracting female companions.[18] The apparent decline in heterosexual activity by modern priests may have as much to do with the decline in real wealth by members of the church hierarchy as it has to do with the Second Lateran Council.[19]

Other interpretations cannot be ruled out, of course. Another plausible explanation is that with the strict observance of celibacy rules by priests and monks, the religious life became less attractive to heterosexual men and thus drew a higher proportion of homosexuals whose lack of interest in marriage could be cloaked in the socially acceptable terms of a religious vocation. We know little about the sexual lives of religious men and women despite the glare of publicity surrounding the sexual abuse of children by Catholic priests.

Are priests reproductive altruists? Or are they cheats, or homosexuals? History provides compelling evidence that many priests cheated on their obligation to refrain from sexual intercourse. In the most scandalous cases, they did exactly what Darwinians would anticipate, exploiting wealth and social power to increase their sexual access to women, occasionally resorting to prostitution and rape. Some may have used their privileged access to women, as priests, to conduct clandestine affairs. According to Boccacio's

Decameron, written in the fourteenth century, the priest was considered an ideal choice as a woman's extramarital partner because he could make house calls at any time of the day without exciting suspicion.[20] Such insinuations have little objective backing, however. Whether priests actually exploited this window of opportunity is unknown. An understanding of the Darwinian basis of human sexuality suggests that many ordinary men would. Priests may not be ordinary in this respect.

Every profession has its cheats and rotten apples, and the religious are no exception. Despite this, it seems reasonable to conclude that most modern Catholic priests and monks do not have sex with women on a regular basis. For some, this is no sacrifice because they are homosexuals. Sources inside the church agree that many modern priests are homosexual, but estimates of the numbers diverge widely. According to Rev. Donald B. Cozzens in *The Changing Face of the Priesthood*,[21] estimates range from 23 percent to 58 percent. Assuming that the truth is somewhere close to an average 40 percent, most priests are likely to be heterosexuals and thus renounce the natural satisfaction of sexual desire through intercourse with women. To the extent that this form of self-denial promotes the interests of the religious community that they serve, it is altruistic. Celibacy is potentially beneficial to the Catholic Church because it prevents the economic resources of the church being diverted to offspring of clergy and other religious professionals. Priests may also be able to contribute more work hours to the church because of the absence of competing family obligations.

Some scholars would dispute the practical importance of these putative benefits of priestly celibacy, pointing to the large, and growing, number of established churches that permit marriage for ministers. There is little evidence that celibate clergy are actually more religious than their married counterparts.[22] Regardless of the validity of these arguments, there is little question that most celibate clergy make a personal sacrifice that may contribute to the vitality of the church and the welfare of its members.

Such reciprocal altruism is not very different, in principle, from the food sharing of ancestral hunters. Instead of giving up food to others, the priest or monk sacrifices sexual intercourse. This is very bad for their individual reproductive success, but the emotional structures on which the sacrifice is based arguably allowed our ancestors to work well in groups and to benefit from mutualism in the uncertain business of hunting. Just as a good hunter is given high social status in recognition of his contributions to group welfare, so a dedicated priest is rewarded by respect and the material prerogatives of status that are psychologically satisfying but do not increase their reproductive opportunities, making them reproductive altruists analogous to worker bees. Priests have generally enjoyed a comfortable life, and the sexual sacrifice that went with service to a congregation was thus repaid in other ways. The communal benefits of a priesthood are difficult to quantify but they could be substantial if religious observance improves psychological well-being and health, as recent research indicates. What about the sexual sacrifices of nuns?

SPARE WOMEN AND CONVENTS

Women often enjoy a special status as noncombatants in wars, suggesting that the survival of women is more critical for warlike families, tribes, or states than the survival of warriors. Women are perceived as the precious vessels containing future generations. Inclusive fitness can be protected if daughters and granddaughters survive, even though a large proportion of males are killed in battle. The reproductive expendability of men is a side effect of the fact that a small number of men are capable of fertilizing a large number of women. Warlike societies are frequently polygynous, with one man doing the reproductive duty of several in a monogamous society. If mortality in war is randomly distributed, the loss of males has no effect on the inclusive fitness of families because their surviving sons sire the same aggregate number of children by taking more wives.

Despite the special solicitude for women's lives in most societies, there are some in which women are not valued. In general, these are the ones in which many women have little prospect of raising children successfully due to the peculiarities of the marriage market and local economy. In some such situations, female children are killed at birth or ill treated during childhood. In others, "excess" women are siphoned off into religious communities where they are no longer a drain on the family economy.

Leaders of the British colony in India in the early nineteenth century were appalled to discover that high-caste groups such as the Jharega Rajputs of Kathiawar in Gujarat, a region of Western India, practiced systematic female infanticide. In some clans, no female children survived. Even after half a century of determined British efforts to root out the practice, it still continued, with over half the female infants being killed in some groups.[23]

As incomprehensible and heinous as such a war on females might seem, the Rajputs themselves had a clear and simple explanation: their daughters were being disposed of because they had little prospects of marriage and were thus destined for a life of isolation and misery. Behind this simple explanation lay the complex workings of the caste system, which dictated that women should marry men of equal, or preferably higher, status than themselves. Due to the practice of females marrying up the social pyramid, there was a great scarcity of husbands at the top. The few men of the highest castes were competed for by a large pool of women of lower castes. Because they were in such demand, high-caste men could expect families of prospective brides to pay huge dowries to establish their future descendants in wealthy social circles. Not only was it extremely difficult for high-caste women to find husbands, but the financial requirements for marrying them off was ruinous to their families.

There are many other examples of societies in which parents prefer sons to daughters, often with tragic consequences for female infants and children. A contemporary example concerns the practice of selective abortion that increases the number of male children born in India, China, and other Asian countries. The problem has been exacerbated by China's one-child policy. If Chinese parents are going to have only one child, they desperately want it to be a son. The reasons are complex, having to do with differing sex roles. Men still earn more than women and have greater responsibility for aged parents. It seems strangely inconsistent that adult women are generally favored, particularly in respect to warfare, whereas males are strongly preferred at earlier ages. There are also societies where adult women are devalued due to a difficult marriage market. When there is a large percentage of females in the population, and when polygynous marriage is not allowed, many women fail to marry. In the past, some of these single women discovered religious vocations and spent their lives locked away in convents removed from the world of marriage and children.

A striking example of "excess" females in the population of Europe during the late Middle Ages is described by the late sociologist Marcia Guttentag and her social psychologist husband Paul Secord in their book *Too Many Women*.[24] The sex ratio imbalance was due to the confluence of several factors. Many men died in protracted wars between European powers like France and England. What is more, this was a period of comparatively rapid urbanization, and cities often had more women than men in their populations. One reason is that urban conditions are less difficult for women than the harsh drudgery of peasant life. Men were better suited to the heavy work of farming, so medieval families in difficult times favored male children over females. Boys were better nourished and more likely to survive to maturity. Given that sons were both better equipped for heavy work and more likely to remain with their families of origin, they repaid the favoritism they had enjoyed during childhood by supporting parents in their old age.

Despite the better prospects of male peasant children, the late medieval period saw a large decline in the male population. For one thing, there was a huge increase in the urban population as modern cities developed and attracted peasant migrants. Plague and other disease epidemics periodically decimated the new cities. Men were more vulnerable to these scourges because they often remained in cities to work at trades and business, sending their wives and children to the country to escape the worst epidemics. Male mortality in war was also very high because of the prolonged nature of medieval wars between European powers.

High male mortality made it very difficult for women to find husbands. Some indication of the scope of the problem can be gleaned from the structure of urban populations in the early fifteenth century, as reflected in tax records. The proportion of men in various European cities ranged from 92 per 100 women to only 83 per 100 women. Given that approximately 106 male babies are born for every 100 females, this means that male mortality in Europe's medieval cities was about 18 percent higher than female mortality.[25]

For many women marriage prospects were dismal. Wealthy families often bought their way out of the problem by providing their daughters with substantial dowries that made them attractive to potential husbands. As in the case of the high-caste Indians, the dowry system began to get out of hand with desperate parents offering ever larger sums to marry off their daughters. Some cities, like Venice, tried to solve the problem through legislation. A 1420 law capped the maximum dowry permissible by law at sixteen hundred ducats, a princely sum. This law was more honored in the breach than in the observance. Another law, passed in the early 1500s, raised the cap to three thousand ducats. This not only undercut the point of the earlier law but also publicized the fact that the earlier law had not been observed.

In addition to providing hefty dowries, some affluent parents tried to get a jump on the competition by marrying their daughters off at a very early age. Child marriage has rarely worked out well, and some young women escaped from objectionable marriages by discovering religious vocations. Others entered religious life after their elderly husbands had died. Among the prominent women of the period, both Saint Rita of Cascia and Saint Clare of Pisa were married at the age of twelve years, perhaps as much as five years before reaching puberty, given the much later age of sexual maturation then than now.

The hordes of unmarried, and unmarriageable, women in Europe, beginning in the late twelfth century, posed a formidable social problem. In the earlier decades of this crisis, many young women were siphoned off by convents. They resolved their difficulty in worldly marriage by becoming brides of Christ. This was only a temporary solution, however. Available spaces in existing nunneries were quickly filling up.

In those days, convents were attached to male religious orders who saw them largely as irrelevant and an inconvenience. As more women joined up, the orders became alarmed at the magnitude of their responsibilities. In 1228, the Cistercian order decided to call a halt. They issued a statute banning the creation of new convents. They even put a damper on existing nunneries by withdrawing pastoral care from priests.

The religious orders were frankly antifeminist. They believed that women were incapable of the kind of self-discipline expected from monks and priests. Women's irrationality and inconstancy supposedly left them prey to heresies, fads, and prophecies. The imputation of spiritual inferiority may explain why so many of the religious ascetics and saints of the medieval period were actually women. They had to go the extra mile to prove themselves.

Unable to marry and shut out of the main alternative career—religious life—women created their own communities that were halfway between convents and completely secular communities. They were known as Beguines. This word is a variant of *Albigensian*, which denotes a heretical sect that rejected marriage and reproduction and even hated children. Unlike nuns, many of whom derived from affluent families, some Beguines earned money doing manual work outside the community. Some begged. Others worked tirelessly in the service of the poor and needy. Most Beguine women were of humble origins, but some were wealthy enough to live on the income from their personal fortunes.[26]

The Beguines were religious in an everyday sense, with an emphasis on prayer and Christian living. On the other hand, entrants did not take a vow of chastity, and they were not permanently committed to the communal life. Some viewed the Beguines as a stepping stone between their families of origin and marriage. Those that did not leave to marry occasionally left to found independent households.

The Beguine communities were hated by the Vatican as quasi-religious entities that fell outside the direct line of church authority. Despite official attempts to get rid of them, Beguine communities flourished and grew into an important social institution that protected single women from the hostile social environment outside their walls. Thus, by the end of the fourteenth century, 6 percent of Frankfurt's women lived in Beguine communities.[27]

A much larger proportion of single women lived in independent households. As many as a quarter of Frankfurt's households were run by independent women, according to tax records. This gives one some idea of the difficulty of getting married, if you consider that independent living was not viewed as a desirable state. In those days, marriage was the only desirable living arrangement for single women unless they remained with their families of origin or entered religious life. The existence of the Beguines as a parallel institution to convents is of considerable interest from the perspective of reproductive altruism. It is fairly clear that the Beguines did not wish to forgo reproduction. Instead, they entered conventlike institutions because they encountered a difficult marriage market, could not marry, and had nowhere else to live. If they were forced by circumstances to postpone reproduction and if they did so contrary to voluntary choice, they can hardly be considered reproductive altruists. If not, then what about nuns?

ARE NUNS ALTRUISTS?

Even the most sympathetic historians have never claimed that the Catholic Church was indifferent to money. From time to time, it has weathered such scandals as the sale of religious indulgences. This is the ultimate blasphemy in a church whose founder went to the trouble of personally kicking the money-changers out of the temple in Jerusalem, a singularly militant action in a life otherwise devoted to nonviolence. Convents have also taken a businesslike approach to recruitment. Thus, many of the medieval nuns were from well-off families. Upon entry, they were expected to contribute some of their worldly wealth to the community. This meant that daughters of the richest homes could aspire to joining the classiest convents.

Parents contributed to the convents that took their "excess" daughters off their hands. This suggests that, apart from their charitable works, convents were providing an important service to the larger community by relieving them of unmarried women who contributed little to family economics. From a reproductive perspective, convents performed the same function in European countries as infanticide in India: they relieved families of the burden of unmarriageable women.

Even though convents performed a useful nonreligious function of supporting single women throughout their lives, it would not be true to say that individual nuns behaved altruistically. They did not forgo reproduction in order to make a contribution to the outside community. First, they had little prospect of marriage, either because they were personally ill-disposed toward marriage or did not attract a spouse, they encountered a scarcity of single men of appropriate age, or their parents were unwilling to provide a large enough dowry to make them competitive in the marriage market. If they would not marry anyway, joining a convent did not affect their reproductive prospects significantly given that historical ratios of single parenthood were very low (typically varying from about 1 percent to 10 percent).[28] Second, being cloistered meant that nuns were, by definition, cut off from the larger community and therefore physically incapable of contributing materially to others. Whatever else one says about the spiritual functions of convents, it is fairly clear that nuns are not reproductive altruists who refrain from having families in order to help their local communities.

It may seem like a perverse form of sexual discrimination to claim that nuns are not altruists, in the narrow evolutionary sense of reproductive altruism, while priests might be. Yet, roughly the same arguments apply to monks. There are some significant differences. Men of the past would have found it much easier to make an independent living, so that monasteries could not expect to receive an equivalent of the "dowry" payments made on behalf of brides of Christ. As we have seen, monks of some historical periods, particularly the late Middle Ages (i.e., after 1000), were far from being committed to chastity, and many had concubines and children, particularly those that had the necessary wealth and social status. There have been far fewer scandals of unchastity in the case of nuns.

As far as religious chastity in general is concerned, a reasonable case can be made that priests might be reproductive altruists. This case is weakened, however, by high rates of sexual activity among some medieval priests. It is also weakened by the very high rates of homosexuality among modern priests. For a homosexual, who is unlikely to father a child anyway, not only is renouncing marriage no personal sacrifice, but it is not going to have much effect on personal reproductive success. Monks and priests are not the only ones to renounce sexuality and reproduction. Chastity was a feature of the lives of the Beguines, for example. It was also adopted by the Shaker community. Throughout history some women have rejected marriage, viewing it as a form of reproductive slavery and have embraced celibacy as a way of avoiding feminine reproductive roles.[29] Such women are clearly in the minority and their behavior is challenging to a Darwinian analysis, although some may have invested in close relatives. Even some modern feminists have adopted chastity as a means of asserting control over their lives. Do any of these nonreligious celibates qualify as reproductive altruists?

NONRELIGIOUS CELIBACY

Religious celibacy may or may not qualify as an example of reproductive altruism. There are many forms of secular celibacy that are also worth considering as possible examples of altruism. As already mentioned, women who fail to reproduce frequently devote their lives to raising the children of close relatives. This widespread human pattern is similar to the helpers-at-the-nest phenomenon in which juvenile birds help to raise their siblings. Such helping is easily explained as an example of kin selection because helpers succeed in transmitting copies of their genes into future generations indirectly via the reproduction of close relatives.

Males as well as females give up, or postpone, personal reproduction in order to help relatives. This could be true of humans as well as other species. A possible example of male helpers at the nest concerns German peasants of the Krummhorn region in the eighteenth and nineteenth centuries, studied by German anthropologist Eckart Voland and others.[30] Land holdings were passed from generation to generation through the youngest son, an unusual pattern of inheritance that helps to keep land holdings in the family. Older sons typically refrained from marriage; either they emigrated to towns, or they remained at home to work the farm, thus contributing to the household economy of their married brother.

This is a fascinating example of voluntary celibacy because the landholders were at the top of the social ladder. Some were quite wealthy. Yet, they refrained from marrying down, even though their social rank would have made them attractive to daughters of small holders or daughters of landless laborers. Refusal to marry down meant that land holdings could be maintained intact, which contributed to the perpetuation of the family in the same place, and hence to genetic survival. More than one-third of the sons of farmers emigrated because there was no new land to be obtained in the area which had been fully settled for centuries. Almost one-fifth (18.8 percent) remained at home as bachelors, contributing to the support of nieces and nephews through their farm work. Being unmarried in that social context, they are presumed to have lived out their lives as celibates and to have had no children of their own. For the entire Krummhorn community, including small holders and landless workers as well as farmers, 13.5 percent of men born in the region remained celibate, compared to 9.3 percent of women. These numbers underestimate the true incidence of celibacy in the Krummhorn agricultural community because approximately one-third of those born in the region emigrated, and their reproductive history is unknown.[31]

The apparent reproductive altruism of men in the Krummhorn bears analogy with the marriage-resistance movement in southern China in the nineteenth century among female silk workers.[32] Marriage resisters took a public vow to remain unmarried and to refrain from sexual relations with men. This vow established them as independent adults for whom parents had no responsibility to arrange a marriage. It also entitled the women to be remembered after death through ancestor worship in their homes of origin. (Married women were normally remembered in the homes of their husbands and single women were not commemorated at all.)

The female silk workers, who were comparatively well paid, organized themselves into cooperative dwellings of six or eight individuals, establishing funds to be used for emergencies, holiday celebrations, retirement, and even funerals. Women in these cooperatives often formed sexual relationships with each other. They lived either as couples or in a menage à trois. Sisterhoods went under names like the Mutual Admiration Society and the Golden Orchid Association. These institutions persisted until the victory of the Red Army in 1949, after which they were banned as remnants of the feudal past.

The origins of the marriage-resistance movement are complex, having a great deal to do with the onerous nature of marriage in Guangdon province. Women married outside their natal village and lived in their husband's home where they were treated like unpaid servants by their mothers-in-law. Economically self-sufficient, the female silk workers did not have to endure these discomforts and indignities.

A number of other factors in the social environment contributed to marriage resistance. For one thing, the custom of single women living together was common. Extended families built girls' houses for their adolescent women. Moreover, customary marriage rules dictated that siblings had to marry according to their birth order. When a young woman joined a sisterhood, her vow of spinsterhood freed younger siblings to marry.

Marriage resisters sent money back to their families of origin, so some parents encouraged daughters to join the sisterhood as one way of increasing their income. The marriage-resistance movement has both similarities and differences to religious convents and the Beguines. Renunciation of marriage and communal self-sufficiency are the most striking similarities. Acceptance of sexual expression among members is the most important difference. Like the Beguines, but unlike nuns, marriage resisters worked outside the home. Unlike the Beguines, who focused on work for charity, the marriage resisters were more strongly motivated to earn money that could be sent to their families. This is reminiscent of the economic behavior of Irish female immigrants in the United States in the second half of the nineteenth century. Many remained single, worked as domestic servants, and remitted their wages to help families in Ireland, many of whom had been decimated by the Great Famine of the 1840s and still lived in wretched poverty. There is little evidence that the Irish women were sexually active, and this would likely have been precluded by the lack of privacy in their lives. In each case, however, the altruism of the working women toward their families of origin looks like kin selection. The Chinese sisterhoods arguably combine elements of straightforward kin-selected altruism with parental manipulation (because parents encouraged daughters to join the sisterhoods and provide economic support for younger siblings).

Given both the birth order rules for marriage and the income obtained from marriage-resisting daughters, it is not hard to see how parents would promote their inclusive fitness by encouraging daughters to join these societies. Even though the daughters did not marry, they contributed to the survival and reproductive success of their siblings and other close relatives. Like worker bees, they refrained from personal reproduction but contributed to the inclusive fitness of relatives, thereby promoting their own inclusive fitness. That the marriage resisters were practicing lesbians is another interesting wrinkle to their story. Their primary inclinations were likely heterosexual, but they found themselves in a social context that ruled out heterosexual relationships. In contrast, a large proportion of modern priests are homosexuals, so that celibacy, defined as renunciation of marriage, is neither a hardship nor a real compromiser of reproductive success. In fact, the modern emphasis on celibacy makes the religious life less attractive to heterosexuals and thus selects more homosexuals. Conversely, in the days of rampant sexuality among some religious leaders during the Middle Ages, there is no evidence of high rates of homosexuality among the clergy.

Heterosexual priests who refrain from sexual intercourse with women could be considered reproductive altruists if their renunciation of heterosexual expression contributed to the welfare and reproductive success of others, whether close relatives or unrelated members of the local community. The Krummhorn men and the Chinese marriage-resistance movement are apparent examples of kin-selected reproductive altruism. Religious celibacy arguably benefits nonrelatives and is a form of reciprocal altruism in which economic support of the clergy by church members is repaid by the psychological benefits of organized religion.

PART 2

Growing Up to Be Good

CHAPTER 4
Why Do People Grow Up to Be Altruists?

Moral capacities emerge for young children in a natural sequence that mirrors their brain development. Knowing how empathy and altruism develop in the life of a child is an important clue to the adaptive significance of kindness for our species.

From their earliest stages of social competence, children reach out to form alliances with caregivers, with older children, and with peers. Their level of altruism is heavily influenced by the context in which they grow up, whether adults scold and punish them or treat them with sensitivity. Children who are constantly criticized develop an expectation that the world is a rather unforgiving place and tend to suspect the intentions of others. Even young children are also quite savvy about detecting when other children take advantage of them (see below).

Children's development of self-awareness is a critical moral milestone that is accompanied by a capacity for embarrassment, pride, and shame: the self-aware moral emotions in humans. Although self-awareness is a crucial feature of human morality and altruism, other species, such as dogs, nonetheless develop moral systems without being self-aware.

SELF-AWARENESS AND ALTRUISM

Dogs are wonderful animals but they are not self-aware. Put a mirror in front of a dog and it will behave as though there is another dog at the opposite side of the glass. At first it may threaten the illusory other dog. Then it may attempt to go around the mirror and confront it. Eventually, though, the dog, being a relatively intelligent animal, despite not being self-aware, will learn to ignore the silent, odorless animal that refuses to come out beyond the pane of the mirror. Animals that are self-aware behave very differently, as we shall see.

Dog breeds differ in their temperaments and behavior, but all are highly social animals. Like humans, canines are group animals with a high level of altruism toward pack members and a propensity for aggression toward strangers.

Domestic dogs are closely related to wolves. The society of wolves has some fascinating parallels with that of humans. Wolf packs are essentially extended families in which the dominant male and the dominant female are usually the only ones to breed. This is rather like human families in which the sexuality of parents is expressed and the sexuality of children is repressed. The status of the dominant animals is constantly being reasserted in greeting ceremonies in which subordinates literally crawl before and abase themselves in front of the alpha pair. Sexual behavior actually occurs in most of the adults in a pack, but the dominant female intimidates her reproductive competitors, making it unlikely that they will conceive or bear live young. The dominant male also prevents other males from mating with the dominant female.[1]

The general warmth of family relations among wolves is remarkable given the reproductive conflicts that exist. Their astonishing level of cooperation and coordination in bringing down large animals like moose are awe-inspiring. Among domestic dogs, huskies are close relatives of the wolf. Their cooperation and effort in pulling a sled over the eleven-hundred-and-fifty-mile route of the Iditarod race from Anchorage to Nome, Alaska, has caused owners to wax lyrical. Dog society is capable of many of the best features of group cooperation that are also found in human societies.

The parallel between the social system of dogs and humans has practical ramifications. One is that the social adaptations of dogs and humans are so similar that dogs can live perfectly happy lives surrounded by humans. It appears that they fit into human societies by treating their owners as top dogs, or dominants. Just as remarkable is the fact that so many human families in wealthy countries choose to keep a dog and treat it very much as a member of their family. Pampered canines receive the best of chow, expensive medical care, and sleep in the homes, or even beds, of their owners.

Why do people lavish so much care on a member of an alien species? A short answer is that on an emotional plane, families do not see the dog as alien. He or she is accepted as an important member of the family. According to John Archer of the University of Central Lancashire, who has conducted a detailed study of dog-human relations from an evolutionary perspective, almost half of English people see their dog as a family member.[2]

Recent research suggests that dogs are particularly good at responding to subtle cues from humans, possibly reflecting an ancient evolutionary association whereby dogs hung around at human campsites acquiring scraps of food and bones. Consistent with this scenario, dogs are extraordinarily sensitive to nonverbal cues from their masters. Research conducted in Budapest found that domestic dogs are far better than wolves at exploiting human cues (pointing, touching, direction of gaze) to find containers that had meat hidden in them.[3] Anyone who eats in the company of a dog is impressed at just how attentive the animal is so that a dropped piece of food rarely hits the ground. Dogs also seem attuned to the emotional state of their masters and can provide appeasement displays when the owner is annoyed, for example.

Successful cooperation between dogs and humans is evidently based on a similarity of emotional adaptations for group living. Given that there are so many similarities, why did humans evolve self-awareness whereas dogs did not? No one really knows the answer to this question, but developmental psychology suggests that self-awareness is linked to human morality and is thus an essential part of the equipment that children use to solve the problems of living in cooperative groups. Individuals who are self-aware can tailor their actions to specific situations. For example, they can decide when to put group interests first and when to look out for number one. Perhaps the reason that people love their dogs so much is that canines are so consistently altruistic in their disposition toward us. Unless you harm a dog, it will nearly always be your friend.

Psychologists refer to such predictable devotion as unconditional love. Dogs express their love for owners in many unmistakable ways: They lick hands and faces in greeting. They wag their tails to express joy and show pleasure when they are patted or caressed. Dogs are never two-faced or duplicitous. Their love is guaranteed. In the company of their pet, people feel relaxed and emotionally secure. There is a measurable drop in their blood pressure, which would explain why pet owners survive heart disease better.[4]

Dogs are incapable of emotional dishonesty because emotional manipulation requires an individual to be self-aware. Individuals who are self-aware can manipulate others by pushing their emotional buttons. A child who is prevented from watching a TV program she likes is perfectly capable of deliberately inflicting a price on the frustrating parent by withholding affection, for example. Dogs are not self-aware so they cannot have an abstract appreciation of their effects on others. As soon as they see their master, they are happy and express their affection. They can no more withhold this affection than a wave can prevent itself from striking the shore. The descriptor "man's best friend" is very apt from this point of view.

People, on the other hand, are intrinsically variable, shifting with their mood and with the political dynamic of the relationship. Being self-aware, humans can project themselves into various scenarios and alter their behavior with a view to maximizing individual gains. The same capacity allows us to experience intelligent empathy for others and prevents us from behaving in ways that would hurt people we care deeply about. Self-awareness not only allows us to manipulate other people, but also permits us to reflect on social relationships in a way that makes us susceptible to the moral emotions of embarrassment, pride, and guilt. According to research by Michael Lewis of the Institute for the Study of Child Development at the Robert Wood Johnson Medical School, children are not capable of these complex moral emotions until they become self-aware, usually by the age of around eighteen months to two years.[5]

SELF-AWARENESS: THE BEGINNINGS OF HUMAN MORALITY

When a baby is around two months old, it thrills parents by focusing its eyes and tracking their movements across the room. Before this point, the infant sees clearly only a narrow sliver of space into which the faces of adults mysteriously appear and disappear. The parents' reaction to the baby focusing its eyes on them is captured in the thought "A person at last!"

First steps and first words are other important milestones in the development of personhood. When these occur, at about eleven and twelve months, respectively, parents have already established a strong relationship to the toddler based on its individuality of temperament, habits, and social responsiveness. Is the toddler aware of itself as an individual?

The standard test of self-awareness for an ape is based on the ability to recognize itself in a mirror. In Gordon Gallup's original test on chimpanzees, the animals were anesthetized before being marked on the brow with white paint.[6] In the human version, the infant's forehead is surreptitiously smeared with rouge. The infant is distracted for a few minutes by allowing him to play with a toy. He is then led into a room where he can observe himself in a mirror.[7]

Children who recognize themselves in the mirror behave exactly the same as the chimpanzees: they use the mirror to investigate the mark, rubbing their finger across it and attempting to remove it. Typically, children younger than eighteen to twenty-four months do not pass the mirror test. By the age of two years, about two-thirds of children show signs of mirror self-recognition. According to the best objective evidence available, younger children are not self-aware.

Michael Lewis, who conducted these tests, notes that the social emotion of embarrassment emerges at the same age as self-awareness. This makes sense if one assumes that the purpose of being self-aware is to make us more effective as social agents. Blushing, averting the gaze, or hiding behind the hands are economical ways of saying, "I know that we have a bargain and I am not living up to my end of it. Even though I have goofed this time, I want you to know that I am conscientious and will try to do better in the future."

The late emergence of self-awareness reflects patterns of brain maturation. The last region of the brain to become fully functional is the frontal lobe. A vital role for the frontal lobe in self-awareness is suggested by the effects of brain injuries in this region.

Phineas Gage, an unfortunate railroad worker of the nineteenth century who had a large iron rod blown through his frontal lobe, experienced a loss of self-awareness that manifested itself in his uncharacteristic use of obscene language. He also reverted to an infantile inability to control his emotions. This suggests that both self-management and social skills are related to the capacity for self-awareness since all of these functions are impaired by compromised function of the frontal lobes.

THE EVOLUTION OF SELF-AWARENESS

The history of psychology is full of blows to human pride. Attributes that we thought of as distinctively human were revealed to be shared by other species. When it was finally recognized that nonhuman mammals not only use tools, but actually make them, one wag declared that humans are the only species that carries its tools in a toolkit. While many people are disturbed by such revelations, animal intelligence need not be seen as diminishing human intelligence. Their gain is our gain. The intelligence of animals such as chimpanzees is not only wonderfully interesting in itself, but also helps us to understand ourselves. This is certainly true of self-awareness.

During the 1960s, Gordon Gallup formally demonstrated that chimpanzees are self-aware in the sense that they can recognize themselves in mirrors in carefully controlled scientific tests. This was hardly a great surprise to anyone who has had much contact with these intelligent animals. Chimps in captivity often use reflective surfaces to examine visually inaccessible parts of their bodies, such as the inside of the mouth, and even to "dress up," where the chimp might place some object, such as a cabbage leaf, on its head and examine the effect.

These responses of chimps to mirrors are very different from the responses of monkeys. The monkeys always see their reflection as another monkey and often persist in threatening it. Self-recognition in mirrors is not "natural" in the sense that it only emerges as a result of prolonged experience with mirrors. We know this from the reactions of stone-age peoples on their first exposure to mirrors. The initial response in a test by Raymond Carpenter of New Guineans was to see the reflections as hostile spirits.

To test his original four chimpanzees, Gallup therefore allowed them to become familiar with the mirrors. He then placed them under anesthetic. While the chimps were out cold, Gallup applied a nontangible dye mark over their brows. What would they do when they woke up and observed their changed appearance in the mirror? Each of the test animals gazed at the dye mark in the mirror, and just like a human, attempted to clean it off using the mirror as a guide.[8] This elegant test, now replicated with scores of chimpanzees, has convinced most reasonable people that mature chimpanzees can recognize themselves in a mirror (infants younger than four cannot), which implies that they have visual self-awareness of a type not seen in most other animals (see fig. 4). While philosophers will continue to claim that chimpanzee self-awareness is of a lower order than that of humans, and animal behaviorists will ingeniously train pigeons to do something that superficially resembles the behavior of the chimpanzees, except that it is patently unintelligent, it is clear that we are not alone in being self-aware.

Orangutans pass the mirror test, but the relatively small-brained gorillas do not, despite being apes, with a single possible exception. The single exception is a "language-trained" gorilla. Apologists for gorillas argue that in gorilla etiquette it is rude to stare. This means that gorillas tend to avert their gaze when they encounter others, unless they happen to be dominant individuals. If they do not look at the mirror image, according to this argument, then they can

Fig. 4. Chimpanzees raised in captivity enjoy putting on a show. They also recognize themselves in mirrors. (Library of Congress)

hardly be expected to learn that it moves exactly with their movements, that it is their reflection. There may be something in this argument for initial exposures, but even using just their peripheral vision, the apes would surely notice that the reflection always moved in the same direction as they did. Following prolonged mirror exposure, gorillas do not show signs of recognizing themselves in the same way that chimps and orangutans do, however. Given their small brains relative to body size, it seems reasonable to conclude that gorillas lack the brain capacity for self-awareness. The single possible exception might be unusually intelligent by virtue of being raised in the laboratory. Another possibility is that researchers have misinterpreted the actions of this single individual, Koko, who is also credited with an unusually large repertoire of signs, compared to other gorillas, and even chimpanzees.[9]

Self-awareness puts great apes in a special category with all sorts of moral and ethical dilemmas about how they should be treated in captivity. The line is further blurred by the fact that human infants are not self-aware, judging from the fact that they do not pass the mirror test until they are two years old.

Why are apes self-aware? What good does it do them in their natural habitat, which is almost devoid of reflective surfaces? Presumably, the great apes, like our human ancestors, could have inspected themselves Narcissus-like in pools of standing water. It is hard to imagine that the ability to recognize themselves in reflective surfaces would have helped them to survive or reproduce, however. Mirror self-recognition is just one manifestation of a more generalized sense of self, although many primatologists passionately dispute this.

The ability to recognize oneself in a mirror requires a sense of self. The capacity to think about oneself would obviously be more valuable to animals having a complex social life because it would allow the individual to make more sophisticated decisions. In general, smaller individuals avoid competing directly with larger ones who can easily overpower them by physical force. This is a simple decision based on fear and the experience of being beaten by larger individuals in early encounters. It requires no self-reflection but can be accounted for in terms of simple rules of animal learning. On the other hand, an individual who is self-aware can solve such problems easily by recruiting an ally. The formation of such alliances is common among chimpanzees. Two unrelated subordinate males can join forces to defeat their common enemy, the dominant male.[10]

Such alliances are also quite common among close relatives in species that lack self-awareness, however. A group of lion brothers may cooperate to take over a pride of females, for example. Self-awareness is not required to explain such cooperation because it may be based on simple mechanisms such as behaving aggressively toward strange males and tolerating familiar ones that smell like kin.

Common chimpanzees are a highly "political" species capable of forming alliances that promote mutual interests of those cooperating, according to Franz de Waal, who observed a captive colony in Arnhem, Holland.[11] Chimp self-awareness might have evolved to promote reciprocal alliances and other kinds of complex social relationships that require mental simulations of possible interactions and their outcomes.

One problem with this interpretation, however, is that orangutans are also self-aware despite spending most of their lives in relative solitude. Single males defend large territories containing the home ranges of females from which other adult males are excluded. Why are orangutans self-aware? No one has any very compelling idea. One theory, known as the clambering hypothesis, is that as large animals hurtling through the trees high above the ground, orangutans have to be intensely aware of their location in space, which is partly accomplished through vision. Hence the capacity for visual self-awareness.[12]

This is not a compelling explanation because even small arboreal mammals, like squirrels and monkeys, occasionally fall out of trees and injure themselves. Yet, neither monkeys nor squirrels are self-aware as far as we can tell. We do not really know why orangutans have a capacity for self-awareness, but it is still reasonable to assume that having a sense of self is beneficial for social decision-making among humans, and our closest evolutionary relatives, the chimpanzees.

For humans, and possibly for chimps also, self-awareness is connected to moral emotions like guilt and pride. Such emotions promote reciprocal alliances by helping individuals to live up to their obligations in an alliance. Yet it does not follow that self-awareness is necessary for reciprocal altruism to occur. After all, look at the reciprocity of vampire bats, which are rather unlikely to be self-aware. This raises the question of whether morality is unique to human beings and other self-aware animals. Can you have reciprocal altruism without having moral emotions to keep the system balanced or fair? Are other species capable of morality?

DO ANIMALS HAVE A MORAL SENSE?

Some animals are surprisingly capable of keeping up their end in a bargain. This is true, for example, of the large reef fish that benefit from the services of the cleaner wrasse, a small, brightly colored reef fish. The cleaner fish operate cleaning stations on coral reefs where client fish line up like men at a barbershop. The wrasse swim inside the mouths and gills of their clients, eating off their parasites.

Once their skin parasites have been removed, nothing would be simpler for the client fish than to make a meal of their benefactor. Yet, they take the high road, never preying on the cleaner fish. The cleaners also behave "honorably," sticking to parasites and never attacking the soft tissues of the fish itself. When the client fish are being cleaned, they go into a state of deep relaxation that makes them particularly vulnerable to attack. This reciprocated altruism is made all the more remarkable by the existence of false cleaners—mimics of the cleaner wrasse—that are tolerated by their clients and dart in to take a bite out of their gills.[13]

Just because the fish live up to an implicit bargain does not mean that they have a moral system as such. We can imagine a much simpler way in which the cleaners and their clients operate. Thus, the cleaners maintain territories where particular fish show up when they need to be cleaned. During busy periods, large fish have even been observed lining up for the attentions of the cleaner. While fish are cleaned they enter a drowsy state, making it unlikely that they will begin feeding. True cleaners are specialized for feeding on fish parasites and may prefer such prey to chowing down on the vulnerable client fish itself. It is thus possible to imagine that the evolution of cleaner wrasse and their client species proceeds in tandem so that the cleaner prefers to eat parasites and the client refrains from eating the little animal that provides it with such an essential service. There is no need to imagine that either the cleaner or the cleaned is susceptible to moral emotions or moral imperatives. Even if they felt absolutely no obligation to each other, their mutually beneficial relationship could be maintained by natural selection operating on the habits of each species.

Real cleaners refrain from biting their clients for the same reason that barbers avoid cutting their patrons: it would be bad for business. Snap-happy cleaner wrasse would lose their clients to more discriminating cleaners. The different fish species live up to their bargains without having any understanding of what the bargain is.

Fish are essentially reflex machines whose reciprocal altruism may not require any keeping of the score. When clients are bothered by the accumulation of parasites on their skin, they simply take their place at the cleaning station. After they have been cleaned they depart. Their actions may be accounted for in terms of simple learning principles, specifically reinforcement. In reinforcement, whatever action produces desirable effects, whether it is a person wearing a coat on a cold day or a hungry rat pressing a lever to receive food pellets, gets strengthened and is more likely to be repeated.

The mutualistic relationship between vampire bat friends is considerably more complex than the cleaner–client fish interaction for several reasons. First, as mammals, bats have a richer emotional life. This can be inferred from comparative anatomy. Mammals show elaboration of structures below the cerebral cortex, known collectively as the limbic system, which play a role in emotional experience. One reason for the development of mammalian emotions is the much greater amount of parental care seen in mammals compared to reptiles, their immediate evolutionary ancestors. Mammals are also better equipped for social cooperation, particularly within families. Some of the most remarkable feats of such cooperation are found among cooperative hunting species, as we've seen in wolves and hunting dogs.

Another reason that we can suspect bats of having more prospects for moral behavior than fish is that their cooperation is based on stable relationships between pairs of friends who recognize each other based on their calls. Bats evidently build up relationships of mutual dependency that involve expectations of giving and receiving help. Moreover, if a bat refuses to feed a starving companion, she can be driven off by her roost companions, which means that if she fails to feed, she is liable to die of starvation. Such ostracism looks like moral behavior.

Whether nonhuman animals are generally capable of moral behavior is controversial. Psychologists have designed experiments in search of an answer.

Can animals learn to obey rules that go against their natural tendencies? Psychologist Hank Davis of the University of Guelph, Ontario, conducted an amusing sequence of experiments to establish the answer for rats. In his procedure, a rat located food at the other end of a straight runway but was allowed to eat only some of the food. The rats were surprisingly good at learning the rule and successfully inhibiting their tendency to eat after they had consumed the permitted amount. What would happen if the experimenter, who punished wrong behavior, was not present? The rat immediately ate all of the food, prompting Davis to conclude that rats are not capable of a high level of moral behavior after all.[14]

Conceptually similar experiments were carried out by early experimenters using dogs as subjects. The dogs were punished for eating a tasty snack: they got a thwack on the snout with a rolled-up newspaper. Dogs who were punished immediately refused to eat the tempting food again even though they were given fifteen-minute opportunities every day for three weeks by the sadistic experimenter. Delayed punishment was not nearly as effective. This crude but interesting experiment suggests that even intelligent animals like dogs would have trouble learning to follow moral rules that were not backed up by punishment systems. The problem is that when punishment is delayed, the dog has trouble learning what it has done wrong.[15]

Despite their learning limitations and lack of self-awareness, canids, specifically wolves, manifest an effective moral system in their natural habitat. While wolves are pack animals, hunting more effectively in groups, only the dominant pair is allowed to breed, as noted above. For other adult females, reproduction is prevented by harassment from the alpha female.

Among adult males, an interesting moral test occurs when the alpha female comes in heat. Can they inhibit their sexual impulses, or will they transgress against the moral code of wolf society? The answer is reminiscent of human adultery—it doesn't happen very often but it happens nevertheless. It is difficult to say whether the sex drive of nonreproducing males is inhibited by fear, a phenomenon that occurs for human males, for example, or whether they are attracted to the alpha female but more strongly motivated by fear of the alpha male.

Interestingly, just about all adult males and females in the pack are sexually attracted to each other. Nonbreeding females come into heat and solicit matings from the males in the pack, but they may not actually copulate unless the dominant female dies or loses her status. Suppression of reproduction among subordinate wolves is fascinating because it is entirely behavioral. There is no physiological reason that all of the wolves could not reproduce. Their reproductive altruism is thus very different from that of worker bees, which are sexually immature and physiologically incapable of reproduction.[16]

Alpha males are very likely to sire the single litter of pups that the pack rears each year. In addition to intimidating rivals, wolves (like domestic dogs and other canids) have a coital lock that holds the penis inside the female after insemination. This mechanism helps ensure that other males do not copulate with the female around the time that she is ovulating.[17]

A subordinate male wolf occasionally succumbs to sexual temptation and copulates with the alpha female, who is not only sexually attractive and sexually receptive but may actively court him. In at least one instance, the group under study cooperated to drive out the offending individual.[18] Banishment is a great price to pay given the difficulty for a lone wolf, or even a pair, in bringing down enough prey to survive. A lone male wandering through the territory of a wolf pack will be aggressively driven off or even killed, since he competes with the residents for limited game.

Wolf society thus provides an interesting possible example of a moral system. It contains the two critical ingredients, namely a system of behavioral rules that goes against individual impulses and a system of group punishment against transgressors. As noted earlier, similar enforcement may occur among chimpanzees.[19]

An enforcement system for transgressors of a social role is not peculiar to wolves or chimpanzees, however. In the small social circle of reciprocating vampire bats, an individual who has recently fed but refuses to feed a starving companion is also liable to be kicked out of the group. As in the wolf case, failing to put the interests of the group before immediate individual interests can have devastating consequences for survival.

Nonhuman animals can evidently follow socially imposed moral rules. How these rules are established is an unsolved problem. We know that the reciprocity arrangement of bats is maintained by natural selection because of the survival advantages of individuals practicing it. Reproductive altruism among wolves is probably maintained by kin selection because more than two hunters are necessary to maintain a successful breeding unit. How the rules are acquired in the life of an individual is an unsolved mystery.

However the rules are acquired, moral systems demand a high level of group solidarity. Therefore, these rules are found in extremely cohesive societies such as those of wolves. In such societies, the social environment somehow controls individuals, causing them to act contrary to their individual inclinations. This is the essence of morality. Human parents often expend a lot of effort in trying to get young children accustomed to sharing with siblings and peers. These attempts are a major part of socialization, the process through which children learn to participate in reciprocity networks that are such a conspicuous and important feature of human societies.

THE EMERGENCE OF HUMAN ALTRUISM

However animals accomplish it, the impulses of the individual in respect to food and sex, for example, can be brought into line with the needs of the group. We know a great deal more about human socialization, thanks to thousands of studies by developmental psychologists. These paint a complex picture of human altruism as a product of genes as well as the environment and as affected by families as well as local communities. Perhaps the most striking conclusion to draw from the emergence of altruism during childhood is that it is a normal feature of interactions between children and others. What is more, reciprocity, the making of bargains, emerges naturally, suggesting that evolution has endowed our species with attributes that promote reciprocal altruism, or give and take.

In early life, as everyone knows, children are entirely selfish. Their lack of altruism has a very simple origin. For them the world revolves around their needs and comfort. This makes sense for creatures that are helpless to care for themselves and must rely on the kindness of parents and other people for food, shelter, and protection from danger. Even so, the stirrings of altruism occur surprisingly early. By the age of twelve months, some children recruit the interest of others in an interesting event by pointing. It is as though they wish to share an entertaining perceptual experience with the caregiver. At the same age, they sometimes offer a toy to a companion, although it is unclear whether the motivation is altruistic. It is possible that they are merely curious as to how the toy will look in the hands of another. Either way, what looks very like altruism occurs at a remarkably early stage in development, suggesting that altruistic behavior of our species may not require a high level of intelligence.[20]

Nevertheless, human altruism clearly develops in synchrony with intelligence. By the age of two years, most children have the brain capacity for self-awareness and show unmistakable signs of moral emotions like embarrassment and pride. By the age of three, their propensity for reciprocal altruism has reached a surprising degree of maturity. Researchers find that children no longer share toys indiscriminately. They are much more likely to offer a toy to another child if that child has previously given them a plaything. What is more, if the other child had previously hogged all of the toys, children as young as two-and-a-half years will hoard toys and refuse to give one to the selfish child, who is left with nothing to play with. Even at this tender age, there are signs of moralistic aggression, which is an indispensable element for preserving systems of reciprocal altruism whether they are found among bats or humans.

There is little doubt that this altruistic sharing of toys is based, at least partly, on empathy with others. To give a toy to another child who has nothing to play with requires some understanding of, and sympathy with, the plight of being toyless. The phenomenon of one child attempting to comfort another who is distressed provides more direct evidence of empathy. Some two-year-olds are much more likely to comfort a distressed child than others are, and researchers have discovered that emotionally supportive children are more likely to be the ones who are self-aware, as measured by the mirror test. This provides further evidence for the pivotal role of self-awareness in human altruism. It also suggests that altruistic behavior, and the empathic reaction on which it is based, awaits brain maturation. Recall that the frontal lobe, which mediates self-awareness, is the last major region of the brain to mature.

Although children are *capable* of altruistic behavior from an early age, altruistic actions are far less common than selfish ones. Young children find it extremely difficult to share candy, for example. This probably makes adaptive sense because a child's health and survival in the evolutionary past would have been very much determined by his ability to obtain sufficient food. Even during normal times, we can imagine that food would have occasionally been scarce in a hunter-gatherer household. One account of childhood in a hunter-gatherer society by Nisa, a !Kung woman, recalls a constant preoccupation with food, from weaning conflict to raiding the "larder" (food tied to the roof of the family hut) to hunger for meat.[21] There is nothing like scarcity to kill altruism and promote competition. Altruistic children who allowed their hungry siblings to get all of the food would have been ancestors to no one. The logical consequence is that young children today are preoccupied with food and generally averse to parting with food that they enjoy.

Even though young children are often selfish in their behavior, their understanding of altruism may be more advanced than their behavior. It is rather fascinating that even though preschool children do not like to share or give, they are attracted to the idea of helping others. Thus, three-year-olds studied in a kindergarten enjoyed *pretending* to be nice to others in their imaginative games. Five-year-olds generally behave more altruistically but are less interested in playing games where they act out the role of an altruist.[22]

The slow development of a capacity for sharing food is nicely illustrated by a charming early Turkish study by R. Ugurel-Semin[23] of the emergence of altruism in sharing nuts. Children faced the challenge of dividing an odd number of nuts between themselves and another child. They were considered altruistic if they either gave more nuts to the other person or gave an equal number by refusing to allot the odd nut. They were "selfish" if they kept more nuts than they gave to the other child. At the age of five years, only a third of children behaved altruistically: the majority kept most of the nuts. By the age of seven, however, more than two-thirds of the children behaved altruistically.

There are many plausible explanations as to why older children behave more altruistically than younger ones. They may have a better grasp of the situation, including more intuition about what another person might feel or notice. They are also more independent from parents and thus more likely to feel responsibility toward others. Consistent with this explanation, children are more willing to help a toddler, who relies on the help of others, than they are to help an older child.

Children's altruism is very much determined by the situation, which is exactly what we would expect if humans were designed by natural selection to succeed in reciprocal alliances. As already pointed out, even very young children apply the tit-for-tat principle to weed out who should be the beneficiary of their kindness and who should not. If a child has helped one in the past, then one should return the favor when given an opportunity. This simple rule receives a great deal of mileage throughout the lives of adults, extending from trivial gestures like sending a Christmas card, to inviting people to a meal or to stay in one's home. Reciprocity is a governing rule of altruism for humans as well as bats because it works. According to computer models constructed by game theorists, such simple rules of reciprocity work well in stable relationships where individuals encounter each other often, because they can shut out cheats who only want to take and never give.

One context in which children's talents at reciprocity never fail to impress researchers is their ability to play cooperatively. Some leading theorists, like Jean Piaget, were so impressed by the capacity of children to formulate rules of conduct for themselves that they dismissed the importance of parents for children's moral development. Current researchers find that children can be strongly influenced by both parents and peers. By the age of four or five most children engage in reciprocal play, and their games, such as tag, have action-based role reversals. Conformity to the rules is clearly essential for the game to continue.[24]

By the age of seven to ten years, most children participate in games that have a formal set of rules, such as hopscotch, four square, freeze tag, jacks, and marbles. At this age, children actively manage their pretend play, assigning roles at the outset and following a play "script." If the game is not to their liking, they may suspend play to modify the script.

As well as grasping the basic mechanism of give-and-take in their social play, young children are highly discriminating about whom they help in other ways. In general, they have little interest in helping a stranger if they can help a close friend, for example. In one experiment by social psychologist Frederick H. Kanfer of the University of Illinois at Champaign-Urbana, children between the ages of three-and-a-half and six years were given an opportunity to work at a dull chip-sorting task to earn toys for others.[25] Almost none of the children (5 percent) worked for an anonymous child, compared to 55 percent who worked for a toy that would go to a good friend. Children's altruism is thus quite rational. By focusing their good deeds on friends, they strengthen the friendship, which increases the likelihood that their help will be reciprocated in future. Childhood altruism among peers is very much determined by the immediate context, but what about the possibility that some people are just "nicer" than others?

ARE SOME CHILDREN MORE ALTRUISTIC THAN OTHERS?

Personality psychologists have been around the block on the question of whether some people are intrinsically "nicer," or more "prosocial," than others. One way of examining this question is in the negative. Are some people intrinsically nastier than others, destined from conception to lie, cheat, steal, and even murder? Researchers have tackled this question in various ways. One approach is to ask whether people are consistent in their actions from one situation to another. Is the child who steals money from another child's locker also going to cheat on an exam? If a child returns a wallet that they have found in the street, does it mean that they will never copy their homework from another? A classic early study by Hugh Hartshorne and Mark May, published in 1928, yielded surprising results.[26]

This old study is still cited as the state of the art because it remains the most ambitious project of its type. Hartshorne and May investigated the moral character of no fewer than ten thousand children by exposing them to temptations to lie, cheat, or steal in various situations. They found surprisingly little consistency in behavior from one context to another. Children who were honest in a test might cheat in a game on the playground. Those who scrupulously obeyed the rules at home might succumb to the temptation to steal at school. In addition to the lack of consistency in their moral behavior, children's moral views had little impact on their behavior. For example, most children who stole said that stealing was wrong.

Such findings have been quite devastating for personality psychologists. Most now accept that it is naive to expect a high level of consistency from one situation to another. Morality may be defined by the situation as well as the person. Parents who are exacting disciplinarians can expect their children to be unfailingly well behaved at home, but it does not follow that the children will be equally well behaved in a different setting, such as hanging out in a friend's home after school. Children are obviously capable of distinguishing the varying costs and benefits of behaving as they please in different settings.

Yet, it is a stretch to conclude that there is absolutely no consistency in a child's moral behavior. More sophisticated statistical analyses of Hartshorne and May's work have shown that these pioneer researchers were premature in abandoning all hope of consistency. Thus, children who cheat on one test are more likely to cheat on another than a child who was honest the first time. Children's behavior is at least consistent in similar situations.[27]

Sympathetic four-year-olds are more likely to share with or help their peers than their less sympathetic age-mates, according to more recent research.[28] Even young children can thus be consistent across different kinds of moral behavior. Consistency increases with age, moreover, as children's actions are affected by moral thoughts and feelings more than by impulse. "Character" is a work in progress that does not near completion until adulthood.

Behavior geneticists report that altruism, like other personality traits, is genetically heritable, although they have been far more interested in the destructive social impulse of aggression, which is also substantially heritable, as is alienation (or the feeling of detachment from other people).[29] These conclusions are based on paper-and-pencil tests that may have limited relevance for a person's real-world actions, however. Hence the necessity to ask whether actual moral *behavior* is affected by genetic background. The most obvious place that researchers have looked is the genetics of criminal behavior.

Generally speaking, behavior geneticists find that having a criminal biological parent is a significant risk factor for criminality. In one of the most common research designs, researchers compare the degree of similarity between identical twins and fraternal twins (who have the same degree of biological relatedness [50 percent] as ordinary siblings). Twin studies indicate that the heritability of adult crime is quite high, 72 percent based on the average of eight different studies. The largest, and thus most reliable, twin study was conducted in Denmark and involved all of the male twins born on Danish islands between 1811 and 1910. Heritability for crimes against persons was 42 percent compared to 48 percent for crimes against property. Adoption studies that look at the degree of similarity between children and their biological and adoptive parents, respectively, find a somewhat lower heritability, possibly because this research design is better at separating the effects of genetics and environment. Adult criminal behavior may thus be almost as much a function of genetics as it is of environmental influences.[30]

Adolescent delinquency is also genetically heritable according to twin studies. Delinquency is defined by deviant or antisocial acts like cutting school, cheating on a test, or using illegal drugs. As the juvenile version of criminal behavior, it is less legalistically defined than criminality, but researchers assume that delinquency can, if unchecked, mature into adult criminality. Heritability estimates for delinquency range from 30 percent to 48 percent, depending on whether an adolescent is simply classified as delinquent or not or is assigned a score on a delinquency scale based on his self-reported acts of theft, cheating, vandalism, illegal drug use, and so on. (While self-reports are not always highly reliable, they allow researchers to study much behavior that does not come to the attention of police.)[31]

Given that delinquency and criminality may be substantially heritable, the question is, what is inherited? This problem has preoccupied biological psychologists for a long time in the form of a quest for brain differences between hardened criminals who are sociopaths and others. About half of male habitual criminal offenders are sociopaths, or antisocial personalities.[32] Antisocial personality disorder is diagnosed based on a history of childhood conduct disorder, indifference to the suffering of others, cruelty to animals, inability to hold down a job, and difficulty in maintaining intimate relationships.

One interesting early finding was that antisocial personalities are less affected by punishment in experiments. This is not because they are less responsive to pain but rather that they are better able to suppress their own physiological responses to a painful stimulus.

In addition to peculiarities in their physiological responses, which are regulated by the sympathetic branch of the peripheral nervous system, criminals are unusual in the pattern of chemical signaling in their brains. Differences in various brain chemicals have been reported when criminals are compared to others. Most attention has focused on monoamine neurotransmitters, a group that includes dopamine, norepinephrine, epinephrine, and serotonin. These substances are of critical importance in regulating mood and responsiveness to environmental events. (Their activity is increased by euphoriant drugs like cocaine and amphetamines, and reduced by clinical depression.) Monoamine activity is generally reduced in the brains of criminal sociopaths with the exception of dopamine that is elevated in the case of thrill-seekers.[33]

Researchers have found a particularly strong connection between brain serotonin levels and violent crime.[34] This link is so strong, in fact, that if you measure the level of serotonin turnover in a violent criminal's brain (which is done indirectly through analysis of serotonin breakdown products in the blood), you can predict their future criminal violence with greater confidence than by all other methods combined. Violent criminals released from jail have a much greater likelihood of reoffending if their serotonin levels are low. In one study, reoffense was predicted with 84 percent accuracy using serotonin levels alone.[35] Animal research also shows that individuals with low serotonin activity in the brain are more aggressive than others, suggesting that serotonin has a role in aggression for other species too.[36]

Just because there are biological differences between violent criminals and others, it does not necessarily mean that these biological differences were caused by genetic differences. One of the most interesting frontiers in biological psychology is the study of environmental influences on the chemistry of the brain. Research on maternal deprivation in monkeys, for example, demonstrates that individuals raised in social isolation have lower levels of amine neurotransmitter activity in their brains and that they are more prone to impulsive aggression. Human infants are also highly sensitive to the quality of the rearing environment insofar as their development of aggression, and its opposites, empathy and altruism, are concerned. In fact, a history of parental neglect and abuse may in itself be sufficient to push children in the direction of extreme antisocial behavior, even in the absence of any genetic predisposition for antisocial personality.[37]

Monkey research suggests that deprivation of an early relationship with the mother alters brain chemistry. It is reasonable to assume that experiences with parents affect brain function in human children also and that this has consequences for their level of empathy.

LEARNING TO FEEL FOR OTHERS

When we see another person suffer, we are liable to feel distress. Helping the other relieves our empathic distress and thus motivates altruistic behavior. Empathy does not explain all altruistic behavior because people may help others without experiencing a high level of emotional arousal, but it is certainly a feature of our reaction to the suffering of people we care about. From this perspective, empathy can be seen as at least partly an evolutionary contrivance that motivates people to care for their children and other close relatives.

One of the most fascinating, and disturbing, aspects of criminal sociopaths is their lack of empathy, which manifests itself in an astonishing lack of concern for their own children. They may have a genetic basis, but it is hard to believe that these genes would have contributed to reproductive success in the evolutionary past if they were expressed in ancestral environments in the same way as they are in modern ones. Children deprived of support and protection from their fathers in hunter-gatherer societies have a much higher mortality rate in societies like the Ache of Paraguay, for example, where some 45 percent of fatherless children die before the age of fifteen years compared to 20 percent with fathers.

According to psychologist Martin L. Hoffman of New York University, empathy is a universal human trait.[38] This does not mean that all individuals have it, but only that most people in all societies have a biological propensity for developing it if raised in an appropriate way. Sociopaths are the exception that tests the rule. Some may be biologically incapable of caring about others, although this is debatable. Most do not develop empathy for the same reason that socially isolated children cannot speak: they were not raised in an appropriate social environment.

According to Hoffman, children progress through five different stages in the development of a mature ability to appreciate the feelings of others. To some extent this progression is driven by normal brain maturation, but it is also susceptible to profound influence from parents and other caregivers. These stages proceed from an infant crying reflexively in response to another infant crying, to a mature appreciation and empathic arousal because another child is in distress. The more mature type, sympathetic empathy, motivates genuinely altruistic behavior, according to Hoffman.

A baby is distressed at the cry of another infant, and it will cry vigorously in response. This reactive cry is not merely a question of being frightened by a loud sound. Controlled experiments using electronically generated wails of equal loudness are less effective in producing reactive crying. There is also less reaction to the crying of a chimpanzee, or even to the infant's own cry. Evidently, there is something uniquely distressing about the sound of another human infant crying. Hoffman believes that this response is built in as a species-typical adaptation that presumably helped mothers to locate their infants quickly in an emergency situation, such as the appearance of a predator.

The reactive cry is the beginning of empathy because it involves distress in response to the discomfort of another infant. It is still entirely egoistic, though. Emerging on the first day of life, it occurs in a creature that has no real grasp of the existence of other people as separate from itself. The egoism of the reactive distress is seen in the fact that the infant finds the other baby aversive and will often turn its head away to escape the aversive sound. Close to the end of the first year, something closer to empathy is seen. The infant looks sad and puckers up its lips before crying. The cry is sometimes accompanied by quiet whimpering as the young child looks at the other distressed infant. It is as though the infant experiences the other child's distress as emanating from himself. For this reason, the infant will seek comfort from an adult.

Soon after their first birthday, children engage in tentative comforting behavior directed at a distressed peer. Patting and touching soon gives way to more clearly defined attempts at comfort, such as hugging, kissing, providing verbal reassurance, and calling an adult in to help. At this age, children cannot really distinguish between their own needs and those of a friend. Thus a fourteen-month-old boy takes the hand of an upset peer and leads him to his own mother for comfort, even though the boy's mother is present.

This limitation is probably connected to the immature state of self-awareness at this age. After children succeed at recognizing themselves in mirrors, there is a leap not only in their expression of social emotions, like embarrassment, but also in their empathic capacity. With a mature sense of self comes an appreciation that others have inner states such as thoughts and feelings. One illustration of maturing empathy is the case of a two-year-old boy who first gives a friend his own teddy bear for comfort but seeing that it doesn't work rushes off for the friend's teddy bear, which is more effective at calming him.

Such incidents reveal that children are genuinely capable of putting themselves in the other's shoes and feeling the other child's distress. This clearly allows for more effective comforting actions. Children as young as three years are capable of surprisingly sophisticated attempts to comfort others, such as a little girl donating a hat to a friend to compensate the friend for having lost a favorite hat several days earlier.

The final stage in the development of altruism occurs, according to Hoffman, when people are capable of feeling empathic distress that is not bounded by the immediate situation. They are saddened by the unpleasant conditions that they imagine other people as experiencing, whether it is a friend whose parents fight a lot, or children borne by crack-addicted mothers. Instead of simply responding to the sadness of a person we encounter, mature empathy is based on a capacity to build up a holistic picture of what their life is like. By the age of seven years, children recognize that a person's gender and ethnicity are stable and permanent. At the same age, they presumably see other children as having a stable identity and life history. They are thus capable of empathy for the life story of another person. Some developmental psychologists believe that children younger than seven years can respond empathically to events outside the immediate situation, but the precise timing is not very important here.

Children's empathy clearly increases with the development of their brains, thought processes, and emotions. Empathy is not just a product of biological maturation, of course, but is intimately connected to the social environment. Parents are an important influence on the altruistic tendencies of their children. Children are also affected by the social context in which they live, whether this is a large extended family or a nuclear one, for example, and by the degree to which competition, or community, is valued by adults.

EMPATHY AND PARENTS

Parents are often fond of preaching to their children about the value of altruistic behavior. A child is encouraged to share with peers and siblings, which is frequently an uphill battle. Contributions to household chores are encouraged, required, and frequently bribed. Consideration of the feelings of other people at all times and in all places is often exhorted, cajoled, and reinforced.

Despite the persistent efforts of most parents to raise moral children, it is not entirely clear that children pay very much attention to what parents say, unless it is backed up by what the parents themselves do. Everyone knows that if parents want to avoid raising foul-mouthed children, it is no use simply scolding children for their conversational obscenities. Parents must also stop cursing around the children.

The power of imitation extends to altruism as well. Developmental psychologists have conducted many ingenious experiments showing that children will act in a more altruistic way if they see other people doing the same. Kind adults who preach kindness are far more effective in encouraging helping behavior by children than adults who are selfish but preach kindness.

Experimenters have bribed children with candy, money, or toys for their helpful actions. Under such reinforcement, children are more likely to behave in helpful ways, but there is a down side. As soon as the rewards are dropped, the children become less helpful than they had initially been. This is rather like the office brown nose who is excessively solicitous toward superiors, who make salary decisions, but is not nearly so kind to others at work.

In contrast to material rewards, praise, a social reward, is much more effective at promoting kind and thoughtful actions by children. There are two likely reasons for this difference. The first is that praise does not undermine a child's sense of autonomy. The child does not feel that he or she is being controlled by social approval in the same way that money is perceived as controlling, for example. What is more, if a child's altruistic actions are not controlled by material rewards, the child is likely to see them as internally motivated. Children whose altruism is supported by praise are thus likely to see themselves as kind people, but children who get material rewards attribute their helpful behavior to the bribe they receive.

Many parents encourage children to carry out specific acts of consideration for others, whether it is cleaning their rooms, thanking store clerks, or writing letters of gratitude for Christmas gifts. Such habits of behavior exert a lifelong influence on a person's actions. Parents also support the development of empathy in their children by directing their child's attention to what another person is likely to be feeling.

In the real world, parents rarely bribe children for altruistic behavior. Instead, they are quite likely to use power assertion or to withdraw affection when the child misbehaves, neither of which works particularly well either. The point of withdrawing affection is to create insecurity in the child concerning the loss of parental love. Power assertion involves the use of physical restraint and spanking, as well as scathing comments and the withdrawal of privileges. Such techniques backfire, as far as moral development is concerned, because they evoke anger, fear, and seething resentment.

If such techniques are so counterproductive, why do parents use them in apparent attempts to correct their children's behavior? That is a deep mystery. Psychologists have long jumped to the conclusion that power-assertive families are dysfunctional.

Hoffman found that parents who avoid hostility and rely a great deal on careful explanations are more effective at promoting altruistic behavior. He refers to such explanation as *induction*. Parents using induction explain why a behavior is wrong by focusing on its adverse effects for other people. They are clear about how the child must change and may even indicate ways that whatever harm has resulted from the child's actions can be redressed. For example, if the child has made fun of another child's appearance, causing her to be hurt and offended, the parent discusses the psychological impact of repeatedly hearing unkind comments about something over which the victim has no control and may suggest that a compliment be paid concerning the victim's clothes. When parents use such a reasoned and sympathetic approach to moral education, their children develop a much better grasp of moral reasoning, are more empathic, and more altruistic in their actions.[39]

The example provided by parents may be more important than any efforts at moral instruction. Children are impressed by personal sacrifices that their parents are willing to make, for example, such as donating to charity or supporting a political cause. Moreover, the whole tone of the parent-child relationship is critical for moral development. Many developmental psychologists believe that parent-child relationships lay the template for all future social relationships, including their degree of altruism.

One example of adult altruistic behavior is devotion to political activism. Whatever else one says about political activists, they are motivated by a desire to improve the societies in which they live and often incur great costs in the name of the cause, thereby satisfying the biological definition of altruism on which this book is based. When social psychologist David Rosenhahn of Swarthmore College studied activism among the civil rights freedom riders of the 1960s, he was struck by the fact that people varied a great deal in how much they were willing to sacrifice to promote social justice in the Southern states.[40] Some were willing to give up their homes and jobs in order to dedicate themselves to the civil rights movement. Others participated in a freedom ride or two without making substantial changes in the course of their lives.

Rosenhahn discovered that the fully committed activists were often the children of political activists who had sacrificed themselves for causes in an earlier generation. Parents of partially committed activists had preached altruism but had rarely practiced it. Another interesting difference was that fully committed activists recalled warm relationships with parents compared to the more distant and rejecting relationships described by the partially committed activists. Personal example really matters. So does the sort of warm parent-child relationship that facilitates detailed explanation of moral actions.

Children's altruism is deeply affected by the emotional tone of their relationships with parents. This phenomenon is clearly illustrated in a detailed study of verbal development of children whose economic backgrounds varied from welfare homes to working class and professional. Higher up the social ladder, children received more verbal stimulation from their parents and scored higher on measures of intelligence. This is hardly surprising: everyone knows that children of affluent parents experience a more intellectually stimulating early environment than those raised in drab and dangerous slums. They have more toys, more social stimulation, more books, more extracurricular experiences, and more private lessons in music, dance, theater, and sports. Their young brains receive a wide variety of experiences at an age when they can greatly benefit from it.

As well as receiving less verbal stimulation from parents, children in poor homes also experienced a much more negative tone in the parental relationship. The authors of the study, Betty Hart and Todd Risley of the University of Kansas, coded the content of speech addressed by parents to children in terms of whether it was positive (i.e., warm and approving) or negative (i.e., scolding).[41] Children of professional parents received thirty-two positives and five negatives per hour compared to five positives and eleven negatives for children in welfare homes. If the emotional tone of a household is measured as the ratio of positive to negative comments heard by a child, then the emotional tone in poor homes was *twelve times* as negative as that in affluent homes.

It would be surprising if such divergent experiences had no effect on children. As they watched the children play at parenting with their dolls, Hart and Risley were impressed by the way that they had imbibed the interaction patterns of the parents. Children in poor homes scolded their dolls a great deal whereas children in affluent homes offered the dolls a lot of emotional support. According to Hart and Risley, "We seemed to hear their parents speaking. We seemed to see the future of their own children."[42]

Some researchers would dismiss such parallels between parental behavior and social development in children because any similarities might be due to shared genes. Such genetic reduction seems wide of the mark, however. Parental styles show little influence of genetics and are strongly determined by social context, in this case, income level. When parents "pick on" a particular child in a family, that child is more likely to become delinquent than other siblings in the household.[43] It therefore seems naive to argue that correlations between parental treatment and child outcomes are simply due to the propagation of "antisocial genes." Clearly, genes and environment are both important for the development of criminal behavior. Further evidence for the influence of the social environment on altruistic tendencies in children comes from comparison of different societies.

ALTRUISM IN DIFFERENT SOCIETIES

Children in some societies are raised to be far more helpful than others. Beatrice and John Whiting of Harvard University conducted one of the most wide-ranging studies of such differences.[44] They observed altruistic behaviors in children aged three to ten years in six societies: Kenya, Mexico, Philippines, Japan, India, and the United States. The Whitings found that children in nonindustrialized societies were very much more altruistic than children in industrialized ones. They found that 100 percent of Kenyan children scored high on altruistic behavior compared to only 8 percent of American children, for example.

The Whitings interpreted these differences in terms of divergent socialization experiences. In less developed countries, like Kenya, people often live in large extended family groups. Children are economically valued because they help with subsistence activities and childcare. The Whitings deduced that being given responsibilities from an early age, such as farm work or caring for brothers and sisters, encouraged a more altruistic orientation early in life. Children are given so much practice at helping others that it becomes second nature to them. Instead of feeling put upon, they derive considerable satisfaction from being able to contribute to their families in meaningful ways.

Kenya is a fascinating example of a country in which children are highly valued in farming communities. The lack of a legal land tenure system means that if extended families are to hold onto their land, they must produce children to occupy farm dwellings and work the land. Kenyans used to define wealth in terms of their number of children, and with very good reason. The more children that they produce, the more land they can occupy and work, and the stronger the prospects of the family for surviving many generations into the future. The great value that Kenyans placed in their children helps to explain why the Whitings found such a high level of altruism there in the 1970s. As Kenyans have moved into cities in recent decades, desire for children and family size have declined. Presumably, the level of altruism in children has also declined under these circumstances.

When people live in extended families, they spend a lot of time interacting with individuals who are close relatives or who are connected by marriage. By contrast, people in industrialized countries spend much of their time interacting with nonrelatives and strangers. Perhaps for that reason, they are more competitive, emphasizing personal goals over shared ones. People in less developed countries have a more communal orientation in the sense that they play down their individual needs and focus on cooperation and group harmony. The emphasis on competition and individual achievements in industrialized countries may undermine altruistic behavior. Among American four-year-olds, children who were highly competitive had trouble sharing candy, even with their best friends. Less competitive children found it easier to share.

Even in America, there is evidence that children's altruism is influenced by how much is expected of them in terms of helping out at home. Sociologists Scott Coltrane and Michelle Adams of the University of California at Riverside investigated children's household work using data from the Panel Study of Income Dynamics Child Development Supplement.[45] Boys who do housework along with their fathers are more popular with peers and are good at making friends. They make less trouble for teachers and start fewer fights. Children who help with housework are thus more altruistic in their interactions outside the home.

Children are particularly impressed when fathers help out with housework because mothers do more housework in the majority of homes. When the father takes more responsibility for housework, children are impressed with this instance of family cooperation and want to help out themselves, according to Coltrane and Adams. If parents want to strengthen altruistic tendencies in their children, the advice is clear: raise them with kindness and sensitivity and give them opportunities to help around the home.

There is thus some compelling evidence that children's level of altruism responds in predictable ways to the social environment in which they are raised. Kind families produce kind children. Socially integrated communities also foster childhood altruism. The obverse side of the coin is criminal behavior, a most conspicuous failure of altruism. Investigating why crime is common in some societies but rarely occurs in others provides a unique window on the evolved mechanisms of human altruism and selfishness.

CHAPTER 5

Altruism among Thieves

The two key ingredients for a crime to occur are the criminal—a person who lacks empathy toward the victims—and the criminal opportunity—an environment where the benefits of crime exceed its costs. One's immediate environment is probably more important as a determinant of crime than genetics or family background, controversial though this claim may initially seem. Island communities, in which people get to know each other very well and develop mutual obligations, often have exceptionally low crime rates, for example.

The geography of a small island makes certain crimes impractical. It makes little sense to steal cars because there is nowhere to drive them. Theft of household objects is not likely to go unnoticed in a community where everyone knows everyone else's business. Moreover, being detected as a criminal can have very serious consequences. Criminal cheats may be ostracized and denied help even when their lives are in danger. Mathematical theorists are fond of discussing cheating and cooperation in terms of a game known as the prisoner's dilemma.

THE PRISONER'S DILEMMA

Why do some people commit serious crimes while others do not? Criminals are clearly distinguished by genotypes and family environments that reduce their altruistic tendencies and make them more likely to put their selfish interests before the good of the community. In a society where there is little or no crime, genetics and family environment are unimportant because their effects on criminal activity are overwhelmed by the immediate environment. In other words, in a society without crime, it hardly matters that some individuals happen to have a biological predisposition for criminal behavior because that tendency does not get expressed.

Pitcairn Island is one example of a society that has no serious crime.[1] A small remote island in the South Pacific, Pitcairn is inhabited by only about fifty people today. It is an interesting case history concerning the nature versus nurture contributions to crime because most of the population is descended from British felons. These staged a successful mutiny on board the ship *Bounty* in 1789, an event that inspired several popular movies. Although some such depictions are sympathetic to the mutineers and suggest that their behavior was justified by the cruelty of Captain Bligh, their true colors emerged in the orgy of criminal violence that ensued after the actual mutiny (see below).

The same natural experiment has been staged at other times and places. A case in point is the British practice of putting distance between themselves and their social problems by shipping criminals off to distant colonies, particularly those in Australia and New Zealand, which were as far away as it is possible to get from England. Many of the early colonists in Australia were felons. Being deposited on a barren continent without female companionship did not improve the mood of many of these unwilling travelers. Even though crime rates were high during colonial times, however, they are not unusually high today. If crime were genetically determined, such a decline in criminality could hardly occur. The fact that Australian crime rates are no higher than those of any other country with a similar level of economic development provides further intriguing evidence for the capacity of the immediate context to overwhelm genetics and family background as influences in criminal behavior.

There are different ways of thinking about the costs and benefits of crime. One way is in terms of a branch of mathematics known as game theory, which analyzes the fate of different "strategies" (really rules of conduct) when they are applied in a community of individuals who may use different strategies. A simple case involves the analysis of aggression in animals. An example of a strategy is "Threaten all opponents. If they flinch, attack them. If they stand their ground, run away."

Game theorists have devised a prisoner's dilemma game for the specific purpose of analyzing cooperation versus selfishness in social interactions. The game derives its name from a scenario in which two suspects of a crime are being interrogated in separate rooms by the police. Neither knows what the other is going to do and has a choice either of turning in his accomplice or of confessing to the crime himself. In the language of the prisoner's dilemma game, ratting out the accomplice is referred to as "cheating." Refusing to implicate the accomplice is "cooperating."

Imagine a hypothetical scenario in which one suspect cooperates and the other cheats. The cheat gets off scot-free in return for his testimony that is used to convict his accomplice. Given the compelling case that can be made against him using the cheat's testimony, the cooperator receives a heavy prison sentence, say twenty years. If both suspects cheat, both will be found guilty, but they will receive a lighter sentence in return for testifying against the other, say five years. If both suspects cooperate, the police have very little usable evidence and can convict them only on a minor offense that gets both a year in jail.

In this particular example of the prisoner's dilemma, cooperation produces the best result in terms of the total number of years served in prison. When both suspects cheat on their accomplice, they go to jail for five years, but when both cooperate, they only get one year each. From the point of view of the individual, however, cheating is the better strategy, regardless of whether the other suspect cheats or cooperates. If the other cooperates and I cheat, then I am home and dry, no sentence. If the other cheats, then I should cheat also because this will get me a five-year sentence rather than twenty years behind bars. The game thus captures the central problem of altruism, which is that the most desirable solution for the individual is not the same as the most desirable outcome for a group (in this case a group of only two individuals). Hence the dilemma.

The prisoner's dilemma game has a complex history. It was invented by economists John Von Neumann and Oscar Morgenstein in the 1950s and then taken up by political scientist Robert Axelrod who developed a collaboration with William Hamilton, an evolutionary biologist, and his colleague at the University of Michigan. Axelrod, and Hamilton wanted to figure out whether there was any mathematical solution to the conflict between the interests of the individual and the interests of the group.[2]

As a once-off proposition, the prisoner's dilemma game is rather unpredictable. Your choice would be very much determined by a guess as to what the other person would do. If you had both been arrested together on a previous occasion, then the fact of having cooperated in the past would surely be critical information in swaying your current decision. Axelrod and Hamilton decided that a repeated (or iterated) game would be better able to capture the dynamic flavor of altruistic behavior in the real world. They launched a computer tournament by asking sixty-two academics in various fields to submit a computer program that coded for a behavioral strategy, such as always cooperating, cheating if the other cooperated, cheating every third move, and so on. These programs were run against each other in pairs. The programs were then entered into a "second-generation" tournament. Strategies that did the best received the most copies in the second generation to simulate evolution. This process was repeated for many more generations.

One of the most successful strategies was also one of the simplest. Known as *tit-for-tat,* this required an individual to cooperate when he encounters a new partner. From then on, he does whatever the other individual does. If a new person moves to your neighborhood, you begin by being friendly and welcoming. If he responds warmly, you strike up a friendship. If he acts as though he wants to be left alone, you remain virtual strangers. That is the flavor of tit-for-tat.

Tit-for-tat has an advantage in allowing individuals to avoid initial mistrust and begin cooperating quickly. It is vulnerable to cheats but can be exploited only once. Tit-for-tat is also forgiving, responding only to the most recent actions of other individuals and not holding grudges.

Biologists believe they have detected tit-for-tat in all kinds of animal interactions, from grooming in monkeys and impala to the exchange of costly eggs for inexpensive sperm among fish and worms that can produce eggs and sperm simultaneously. Chimpanzees and baboons join forces with individuals who have helped them in social conflicts in the past. Vampire bats refuse to share blood with roost mates who refused them previously.

Tit-for-tat-style reciprocity has been observed among humans in a variety of situations, some of them quite surprising. One of the oddest was described by Robert Axelrod in his analysis of interactions between the British and their German enemies during the deadly trench warfare of World War I[3] (see fig. 5). Remember that soldiers in the trenches were involved in a new and exceptionally dangerous warfare that was a death sentence for the majority who were unfortunate enough to be placed in that situation by officers who were out of their depth, to put it charitably.

Entrenched troops spent their time shelling the enemy, thereby participating in a horrendous and strategically pointless slaughter. Against all orders, the troops on each side developed an etiquette of firing only to the side of enemy positions to minimize the loss of life. Axelrod reports one astonishing memoir of a British officer whose position was unexpectedly shelled by the Germans, although without casualties. Following the attack, a German officer surfaced from his trench and shouted, "We are very sorry about that; we hope no one was hurt. It was not our fault, it was the damned Prussian artillery."[4]

Both sides had realized that with the deadly weapons available to them it was just too dangerous to try and kill each other all the time. It was in their mutual interest to cooperate. Reciprocity had broken out in one of the most unexpected places. There was no

Fig. 5. Trench warfare in World War I. (Library of Congress)

verbal agreement, which would clearly have been treason, but the repetitive nature of trench warfare provided an opportunity for cooperation to occur between the opposing armies.

Social psychologists have carried out many experiments using the prisoner's dilemma game. In a typical game, participants push one button to cooperate and another button to defect. They are told in advance about how their payoffs will be affected by the actions of the other player. If both players push the cooperate button, both get a moderate reward, say fifteen cents. If both defect, they receive a smaller payoff, say ten cents. If one cooperates and the other defects, the cooperator gets only one cent and is thus a sucker, while the cheat gets twenty cents.

People's behavior in such experiments is quite variable, and it illustrates some strengths of the tit-for-tat strategy. Behavior on later trials is very much determined by outcomes of earlier trials. If a person defects on the first trial, it is difficult for him to turn around and cooperate because he has created a hostile environment in which he can expect his partner in the game to defect out of spite. It pays to be nice to people on the first trial to give them an opportunity of cooperating. People who believe that other individuals are likely to be cooperative begin with a cooperative response and have less difficulty falling into a routine of repeated cooperating that is in the best long-term interest of both players.

TIT-FOR-TAT AND CRIME

The conflict between selfish interests and the common good that is at the heart of the prisoner's dilemma game has implications that extend far beyond the pair of individuals playing a game of cooperation and competition in a laboratory experiment. Cooperation and defection describe alternatives faced by people in a wide variety of ethical situations from environmental pollution to crime. In these situations, the individual effectively plays a prisoner's dilemma game opposite his entire community. For example, a person who uses herbicide on his lawn is poisoning not just the undesirable plant species but also the animal life, including worms, larvae, and the birds that feed on them. Some toxins are held in the grass but most get washed into groundwater, ultimately poisoning fish and the people who eat them. Herbicides are legal, of course, and modern chemicals used for this purpose may be less harmful than they used to be, but the fact remains that we would all be much better off if people refrained from using them. This would mean that they would be forced to weed their lawns by hand, or make their peace with invading local plant species. Faced with the smaller personal reward of a dismal-looking lawn, most people prefer to defect on the environment, and on the community, by going for the instant gratification of a weed-free lawn.

Illegal pollution is no different in principle from other kinds of crime. It is true that environmental consequences of pollution are often visible, or scientifically detectable, whereas those of many other crimes are more difficult to detect or quantify. A mugger who extracts $50 from his victim can have a profound psychological impact on that victim and the entire community. Someone who has been mugged may develop a paralyzing fear of going out at night, which not only detracts greatly from his quality of life but could have adverse consequences for his health, prompting expensive visits to doctors, drug treatments, and counseling sessions. Already, the cost of the crime is considerably more than the money that was stolen. Cities that are infested by petty criminals lose business, lose old enterprises, and fail to attract new ones. Street crime is thus an insidious force that attacks prosperity as well as peace of mind.

Crimes do not need to be violent to have such pervasive deleterious effects. Indeed, some scholars believe that the economic cost of white-collar crime is far greater than that of street crime. When the crooked dealings of corporate insiders destroy the profitability, and viability, of huge companies, like Enron, large numbers of people lose their jobs and their retirement funds. For some, this might be an opportunity to start a new and better life. For most, it involves a slide toward poverty and despair. Once again, the costs go far beyond simple economics to the fundamental, but difficult to quantify, issues of human well-being, happiness, and health.

Crime thus resembles a prisoner's dilemma game that pits the individual against the entire community. The criminal is the defector and honest citizens are the cooperators who provide all the advantages of civilized living of which the defector takes advantage. Needless to say, there are other useful ways of thinking about the relationship of criminals to their communities. Any useful theory of crime should help us to understand why crime occurs more often in some settings than others. Crime rates could thus be reduced through environmental and social controls. A workable theory of crime also points the way toward rehabilitation of criminals.

PAYOFFS: THE COSTS AND BENEFITS OF CRIMINAL BEHAVIOR

To obey the law is like cooperating in a prisoner's dilemma, and breaking the law is defecting. Whether people cooperate or defect in this community-wide "game" is determined not just by their individual characteristics—how altruistic they have grown up to be—but also by the payoffs. Some people are undoubtedly more criminal in their inclinations than others, but however selfish they may want to be, in the appropriate community they abide by rules of expected conduct.

Remember that if there is any such thing as a "criminal" genotype, it was well represented among the original European colonizers of Pitcairn Island, murderous thugs whose violent propensities were revealed after they destroyed the *Bounty* and took up residence on the island. It is particularly interesting to look at the Pitcairn Island experience from the point of view of a potential criminal, say a thief. A thief on Pitcairn Island would confront considerable logistical problems due to the closed nature of the community. The island community is closed in the social sense that everyone knows everyone else. This social closure is connected to geographic isolation. Humans are social animals. Marooned on a remote island, Pitcairn's inhabitants are more or less compelled to know each other due to repeated encounters in the few public places where most people show up periodically. Family connections extend far back in time to include several generations of people and their recent ancestors. In all probability, they will extend many generations into the future. This is the mother of all repeated prisoner's dilemmas! Not only do people know each other in a superficial sense, but they are likely to have built up a vivid picture of all the person's actions, private as well as public. We should expect such longstanding relationships and frequent interactions to encourage cooperation. This is clearly demonstrated in connection with the problems of a hypothetical burglar.

Suppose that I am living on Pitcairn Island and I decide to steal my neighbor's clock that is carelessly left standing close to an open window. I reach in the window and place the clock in the voluminous pocket of my overcoat. Stealing the clock was easy, but now what do I do with it? Professional thieves use fences, equally unscrupulous middlemen who accept stolen goods at perhaps a quarter of their true worth and make a quick profit by selling them to more legitimate businesses. Small islands have no criminal industry to speak of and therefore no fences for stolen goods. The best that I can hope for is to wait until some stranger arrives who is in desperate need of a loudly ticking clock to carry back home as a souvenir. For the moment, the chance of profiting from my ill-gotten property is remote. I might as well bring it home and set it up as a companion for the ticking clock in my own cottage. Yet, that is hardly possible because my wife will want to know where I got it. Even if I lie and say that I won it in a raffle, she will inevitably hear that the neighbor's clock has gone missing, put two and two together, and demand that I return it.

Crime is unlikely to pay on a small island community because there is often little that is worth stealing. That which is worth the effort of taking cannot be sold. The thief is compelled to keep it and the victim will inevitably learn of its whereabouts. Because islands are closed communities, what is known to one inevitably becomes known to all. Even if I were single and decided to keep the clock without telling anyone, eventually some inquisitive mail carrier or some curious neighbor child would peep in the window and realize that there was an extra clock. Admittedly, I could keep the clock wrapped up in a sack underneath a loose floorboard, but what would be the use of that? If you can neither sell something nor use it, then it has no value.

Crime cannot be lucrative if practiced against the inhabitants of small islands. What is more, informal punishments can be severe. We often think of the criminal justice system in complex societies as being designed to protect the public from criminals, but it also has the effect of protecting criminals from the wrath of the public, specifically from vigilantes. Pitcairn does not have a problem with vigilante justice, but there is an even more devastating way in which the community punishes wrongdoers, namely ostracism.

Much has been written about the idyllic side of life in small communities like Pitcairn Island. The Pitcairn Island community was founded by nine mutineers together with nineteen individuals from Tahiti, six men, twelve women, and a baby girl. The early years were exceptionally violent: an orgy of homicide wiped out most of the male settlers in the first decade, but not before they had fathered offspring to continue the line. Of the original fifteen men, one died naturally, another evidently killed himself accidentally while drunk, and no fewer than twelve were murdered, presumably reflecting a combination of impulsive violence and the absence of civil authority. The Pitcairn community soon organized itself according to Christian principles, however, and it has survived for some two centuries without any formal policing. During that time there have not been any serious criminal acts. (A 1997 allegation of rape was investigated by a British police inspector and turned out to be groundless.) What could be better than a crime-free community in a paradisal Pacific isle without even a permanent police officer?[5]

There is a dark side to this paradise. Small self-sufficient communities pull together. They exert themselves on behalf of others, working without payment on public buildings, on landing goods from passing ships, on hauling wood for carvings from a distant island. The dark side of this group cohesion is the way in which dissent is handled. To begin with, there is almost no privacy. Islanders spy on each other to ensure that everyone is living up to the group code of conduct. This leads to all manner of small-mindedness, petty jealousies and rivalries, long-running personal antagonisms, and feuds. Islanders not only spend much of their time working for the community, they also sacrifice their individuality and freedom for the sake of the island population as a whole.[6]

Islanders are particularly suspicious of outsiders, sometimes with good reason. English journalist Dea Birkett visited there with the intention of writing a book on the social life of the island. She concealed the fact that she was a writer and even attended church in order to fit in. She then blotted her copybook by having an affair with a married man. The community suddenly became so hostile and threatening that she had to make an unplanned exit on a passing ship. In the course of her ostracism by the community, it emerged that the islanders had never really accepted her, despite their seeming friendliness. They had spied on her personal and professional activities. If Birkett felt betrayed by her contacts, as she reveals in a disillusioned book, *Serpent in Paradise,*[7] the feeling was evidently mutual. Hence the threats that motivated her to leave.

Life in the small closed community run on religious principles has little more freedom than life in a religious cult. Islanders obey the rules but their conformity is obtained through a form of mind control. The needs of the society take precedence over the needs of the individual. Religion obviously plays a role in social conformity but it is not essential. One striking secular example of people putting the community before their individual needs is the nonreligious kibbutzim in Israel.

When crime is viewed as a prisoner's dilemma game, it is clear that most communities are not passive in the face of defection. Rather, they pool their resources to put criminals out of business. In complex societies, identified criminals are taken out of circulation, usually by temporary imprisonment followed by a second chance after the debt to society has been paid through time served. This is tit-for-tat in which the defector is punished but given an opportunity to reform. In simpler societies, there is no prison, as such, but the defector is taken out of circulation through ostracism. He is shunned, or silenced, denied the social and material benefits of the community, and exposed to threats and aggression. Instead of being surrounded by a physical prison, the defector is enclosed by social hostility instead. We can imagine that such discrimination would make survival difficult and marriage impossible. It is also very frightening and depressing. Social isolation is a major psychological stressor that undermines the immune system, making people much more vulnerable to serious illnesses.

There are thus several fairly obvious reasons why living in a small homogenous community reduces criminal defections. Criminal opportunities may have severe economic limitations. The social costs of being identified as a criminal can be quite crushing and disproportionate to the meager benefits. Another major problem, from the point of view of the defector, is the extreme difficulty of escaping detection.

In small communities, everyone knows everyone else's business, making it difficult for criminals to operate. Anonymity is a critical defense for a criminal. Once he or she is identified, the community can easily punish him or her. As cooperators in a prisoner's dilemma, community members want to punish defectors. Merely observing the actions of criminals is a potent method of inhibiting crime.

This phenomenon is illustrated in two modern efforts at crime prevention. One is through architecture. The other is video surveillance in British cities.

THE INHIBITED CRIMINAL

We can all benefit from our mistakes. Architects are no exception. Current interest in the possibility of building houses that deter crime was sparked by a nightmare of crime-infested high-rise public housing in US cities.

The beginning of the end was the dynamiting of the Pruitt-Igoe housing project in St. Louis. Although structurally sound, the development was so overrun by street crime that the housing authority decided to raze it and begin again. This spectacular failure was followed by several others in the vicinity of Newark, New Jersey. The housing authorities were admitting that the flawed architectural design of the buildings fostered crime.

Can buildings really cause crime? Not directly, but they can create an environment in which criminals find it easy to operate. Crime is generally conducted in private to avoid detection or identification of the perpetrators. Any kind of surveillance is bad for criminal enterprise because observers can easily tip police off.

Architect Oscar Newman, writing in 1972, was the first to propose that urban buildings could be designed to prevent crime.[8] Newman's central concept was *defensible space*. This means semipublic spaces around housing developments where residents often spend time and feel a sense of territorial ownership.

Examples of defensible spaces include the steps in front of apartment buildings where residents sometimes congregate as well as playgrounds, picnic areas, and playing fields that ideally separate urban high rises. By occupying these semipublic areas, residents notice the incursion of strangers. Their mere presence makes it less likely that criminal elements will hang around.

Newman's thesis was that some buildings, particularly public housing projects, do not make provision for such defensible spaces.

Lack of defensible space creates an anonymous environment in which crime flourishes. In one Newark housing project lacking defensible space, residents were often attacked in their own hallways. Even when they ran the gauntlet of the ground-floor hallway and entered the elevator, they could expect to meet a hoodlum who demanded payment for allowing them to ride up to their floor. Buildings may not cause criminal behavior but they can provide an environment that enables it to flourish.

Unlike many other visionaries, Newman and his Institute for Design Analysis received an opportunity to test out the defensible-space theory. Consulting with the city of Dayton, Ohio, he supervised the construction in 1992 of fifty-six gates that prevented through traffic in the crime-ridden Five Oaks area. At a modest cost of only $693,000, this project was a real success. In the first year, crime had dropped by a quarter and violent crime had fallen by a half. The idea of the gates was to give residents more of a sense of ownership over their neighborhood. Following installation of the gates, residents began to feel safer and spent more time outdoors. With less danger from through traffic, children were allowed to come outside and play. Due to an increased sense of safety, property values rose and homeowners took better care of their dwellings.

It is difficult to know whether the crime reduction in Five Oaks was caused by a returning sense of community, following the provision of defensible space, or whether fewer criminals visited the region because of the inconvenience of driving around the gates. Newman was involved in another interesting community project relevant to the effect of neighborhoods on crime. The city of Yonkers was required to desegregate its housing. As part of the legal remedy, it agreed to construct two hundred new public housing units in affluent white neighborhoods. White residents were concerned that crime might increase as people from high-crime areas came to live among them. The dreaded increase in crime never happened, however. Rather, the new sense of security experienced by the residents in these enclosed communities encouraged them to establish psychological ownership over the areas around their homes. They erected picket fences, planted flowers, and set up play equipment for their children. They established defensible space around their homes that evidently helped to inhibit crime.

VIDEO SURVEILLANCE AND CRIME

As already noted, communities have an important role in surveillance. In communities with defensible space, citizens not only detect crime and quickly report it to police, but also monitor suspicious characters, alerting police, and thereby even prevent crime.

Being under constant surveillance inhibits crime.[9] Burglars prefer to victimize homes that are poorly kept up, suggesting that occupants are not around very much, than those with neatly trimmed lawns. This is a little irrational in view of the likely correlation between a home being well maintained and containing rich pickings for a burglar. Working without being detected is of critical importance for burglars and other types of career criminals. Once a criminal is seen in action, there is a reasonable probability that he or she may be identified and apprehended.

Although occupants who inhabit defensible spaces inhibit criminal activities, their eyewitness observations are of limited use in actual court cases and their reliability has experienced a continuous assault at the hands of memory researchers. Witnesses to a crime have been found to incorporate details that they read in a newspaper into their memories of the crime, for example. Video surveillance provides many advantages over eyewitness evidence, chief of which is objectivity. Videotapes can be played as evidence in court. This was a mixed blessing in the old days of black-and-white videos because it was often difficult to provide a definitive identification of the criminal due to poor image definition. With the advent of high-resolution color video, criminals caught in the act no longer even bother to contest their conviction for crimes recorded electronically.

Street surveillance systems have been used effectively to reduce crime in the cities of several countries in Europe, particularly England and France. By 1997, England had over 100,000 video cameras trained on busy regions of streets, government buildings, and public housing projects. These cameras are a particularly powerful tool if they are combined with human intelligence because it is possible to zoom in on a suspect while he is committing a crime. The mere knowledge that such surveillance systems are active encourages criminals to move to areas without surveillance. Some security experts believe that in addition to deterring street crime, camera surveillance prevents terrorism. For example, the British security system allows the video images of international terrorists to be matched up with a library of dangerous people so that their walking past a camera on an English street may trigger an alert. Many people object to these cameras, feeling as if "Big Brother" is watching, but given the world we now live in, we must at least consider such options in fighting crime and terrorism.[10]

CRIME IN TOWN AND COUNTRY

Crime is much more of a problem in the modern world in which people congregate in cities than it was just a century ago when most people still lived outside cities. This transition is not merely some fluke of historical change, because crime rates in large cities today are much higher than rates in small towns or among rural residents. In the United States, for example, metropolitan areas have approximately 2.7 times as much violent crime as rural areas (based on the most recent statistics).[11] Americans living in cities of over a million people have annual crime victimization rates of around 43 percent compared to 19 percent for residents of towns with between one thousand and ten thousand people.

At first blush, higher urban crime rates are something of a paradox. After all, when one is in a city, there are usually other people in sight. Why are crimes so common if there are so many potential witnesses? The answer is complex but it can be deduced from the prisoner's dilemma in which the group of cooperators is large and anonymous and the degree of cooperation to be expected is lower. If my next-door neighbor, whom I know well and see often, is robbed, then I am likely to take time off work to go to court and help convict the robber. If a complete stranger is mugged on the street, however, I might be less willing to take the day off work on her behalf. (In some cases, judicial treatment of muggers has been so light that even the *victims* are unwilling to press charges.)

In cities, the criminal benefits from weaker bonds of cooperation due to the anonymous conditions of city life where large numbers of strangers mingle together. Even more important is the fact that the criminal is likely to be unknown to victims. This means that even when the crime is conducted in broad daylight, in the presence of potential witnesses, the criminal is unlikely to be recognized and thus unlikely to be identified as the perpetrator. In their research on crime rates in various regions of the United States, Edward Glaeser of Harvard University and Bruce Sacerdote of Dartmouth College found that the lower probability of arrest and recognition in cities explains approximately one-fifth of the differences in crime rates between large cities and rural areas.[12]

Glaeser and Sacerdote found that the greater monetary benefits attaching to crimes in large cities could account for approximately one-quarter of the differences in crime rates as a function of city size. Wealthy people congregate in cities so that the pickings are rich. There may be nothing worth stealing on a small impoverished island and few opportunities to dispose of it due to the lack of professional fences, but the opposite is true in an affluent city in which there are good economic opportunities for crime and well-developed markets in stolen goods. Criminals congregate in cities in part because that is where the money is.

Urban conditions foster crime for other reasons also, particularly high rates of single parenthood. According to Glaeser and Sacerdote, the biggest factor in accounting for high crime rates in cities is the high number of female-headed households, which can account for over one-third of the elevation in crime rates.[13]

One reason for urban differences in family structure is that large numbers of single women migrate to cities in search of a better life. Because women were more likely than men to abandon family farms, women historically outnumbered men in cities during the period of rapid urbanization that had the effect of "undermining morals," or "liberating women," depending on one's point of view. Extramarital sex was more common and the rate of illegitimacy rose because there was an insufficient supply of marriageable men to satisfy the demand from single women.

Trends in sexual liberation were also affected by the widespread use of effective contraception. With the fear of unwanted pregnancy removed, young women had less reason for postponing sexual relationships. Combining all of the above with increasing earning capacity of women provides some explanation for the rapidly increasing rates of single parenthood among urban populations in developed countries.

Single parenthood is a major risk factor for crime. Thus, historical increases in crime have been strongly correlated with increases in single parenthood. Moreover, anywhere from one-third to one-half of the differences in crime rates between American regions can be explained in terms of differences in rates of single parenthood.

Why might the absence of fathers from homes promote criminality? The answer may be partly economic. Single-parent homes are much poorer than double-parent, dual-income ones even though one partner may work part-time in a low-paying job. The full-time worker is most likely to be the man, and men still earn about 20 percent more than women in the United States and other developed nonsocialist countries. Relative poverty can be important to the development of criminal inclinations for a variety of reasons. Most obvious, perhaps, is the fact that poverty commits people to living in poor neighborhoods that are often crime-ridden and thus a haunt of delinquent youth gangs that provide an entry point into serious criminal activity. Poor homes are also characterized by a more abrasive tone in interpersonal relationships as observed in the conversations addressed by parents to children in Hart and Risley's study of verbal development.[14]

Such differences in the emotional tone of households as a function of social status are consistent with an evolutionary theory of socialization. According to this theory, children raised in a difficult, highly competitive environment in which it is challenging to acquire food and other necessities of life develop a more selfish orientation toward others. They are more likely to look out for themselves. They are less altruistic in their outlook and behavior. They are more suspicious of the motives of others. In the language of the prisoner's dilemma, they are more likely to defect. In concrete terms, they are more likely to carry out criminal actions as a means of pursuing their egoistic needs at the expense of the wider community.[15]

The reasons for higher crime in large cities thus include family structure, and economic deprivation, as well as the criminal opportunities present. Whether cities make criminals, or merely attract them, or both, remains an open question.

Regardless of the role of criminal predispositions, it is probably true that criminals find it more difficult to victimize friends and acquaintances than complete strangers. Every person is part of the social island of their intimate friends and relatives. If they offend against that group, the price is potentially high in terms of direct retaliation, including ostracism. A young man who repeatedly steals money from his parents is liable to be kicked out of his home, for example.

It is emotionally difficult, as well as imprudent, to prey on one's own friends and relatives. One reason that cities are attractive to criminals is that there is a large pool of potential victims who are not members of the criminal's social island.

The argument thus far has been that crime is like a prisoner's dilemma game in which the costs of defecting are greatly reduced by victimizing strangers. When criminals exploit family, friends, or acquaintances, they face severe, immediate, and certain punishment, which is particularly harsh in small cohesive communities where offenders can be shut out of all social benefits, or ostracized. They die a social death and can expect no assistance from others in their hour of need. Some people are more predisposed to repeated criminal behavior than others. Although criminality is partly heritable, environmental influences are stronger so that some communities have no crime regardless of the proclivities of the individuals that compose them. The main requirements for a crime-free society are social closure (each individual is known to all others), interdependence, and an effective counterattack by cooperators against defectors (usually withdrawal of social privileges, like friendship, ostracism at an extreme). If these ideas are correct, then it should be possible to rehabilitate habitual offenders by inducting them into closed societies outside of prisons.

REHABILITATION

The notion of criminal rehabilitation is anathema to many. According to this perspective, the criminal had his chance and he blew it! The prospect of altering criminal behavior by altering the social environment might seem hopelessly naive. After all, crime is a function of individual differences as well as the immediate situation in which the criminal behavior occurs. The leopard cannot change its spots, according to this view, and neither can the criminal. This is a reasonable objection and it underlines the great difficulty of any serious rehabilitation effort. To reclaim criminals, it is not sufficient to change the environment. The environmental change must be potent enough to remake the criminal, thereby reversing the effects of previous environments, particularly that of childhood.

Due to the great social costs of serious crimes, there has not been very much effort to experiment with rehabilitation because of the dangers of reoffending. In the past, criminal authorities have been happy to identify criminals and take them out of circulation in some way, whether by imprisonment, banishment, or execution. By removing defectors, cooperators can stay in business.

Of these three methods of punishment, banishment is the most promising example of a natural experiment in which criminals are sent back into communities. Execution tells us nothing, of course. Imprisonment may be counterproductive when considered as a technique for rehabilitation. Reoffense rates are high following imprisonment, and some criminologists believe that the prison system is essentially a university of crime in which novice criminals and petty offenders benefit from the experience of others to hone their craft or move on to more promising fields in which other inmates are expert, rather like a college student changing majors. A prison is essentially a collection of defectors with a few honest people who were mistakenly imprisoned. It is the opposite of the larger community that contains many cooperators and a few defectors.

One would imagine that life in jail would be very dangerous and violent. In reality, it is not much more dangerous than life in another institution, such as a mental hospital or retirement home. Prisoners live by their own internal codes of behavior that regulate personal loyalty, the exchange of contraband, the granting of sexual favors, and most of the social transactions that are important in life outside jail. They develop cooperative schemes for bribing guards, thereby improving their standard of living.

Prisons are physically closed but they are not closed communities. The constant turnover of players as new people are incarcerated and others are released means that inmates are unlikely to get to know many other prisoners well. Despite this, prisons have many of the ingredients of low-crime communities and have comparatively low internal rates of serious crime even though prison violence is sensationalized in the rare case of prison riots, violent rapes, and homicides.[16] The problem is that the comparatively peaceful community of cooperating defectors inside prisons devote much of their ingenuity to subverting the larger community, frequently including detailed plans for subsequent crimes. Prison inmates constitute an in-group for whom the out-group is the law-abiding community. Friendships formed in jails can be the basis of prolonged criminal cooperation, and they may lead to remarkable acts of altruism.

Depression-era bank robber John Dillinger (see fig. 6) began his criminal career as a very inept robber and spent eight years in jail as a result of his first botched crime. While in jail, he thought about bank robberies and perfected the craft largely by discussing techniques with other inmates. Dillinger was loyal to his jailhouse companions and helped to spring them from their cells. Such acts inspired sincere respect and affection among his criminal companions. The notorious bank robber even inspired trust in honest citizens. Journalist Robert A. Butler was willing to help Dillinger give himself up to police, thereby avoiding bloodshed. Butler was willing to risk dying in a hail of police bullets after only a single meeting with the FBI's most wanted man. The plan failed because Dillinger was frightened off by the accidental appearance of a police car that made him suspect treachery. Within a few weeks the legendary bank robber was shot down by police, just as he had feared.[17]

Loyalty among prisoners can be remarkably generalized as in the case of a former drug dealer known to the author who supervised a construction crew and preferred to hire men who, like himself, had spent time in prison. Interestingly, such people often recognize each other by telltale signs, such as homemade tattoos, frequently on the left hand of the right-handed tattoo artist. Honor among thieves can thus be literally true.

Prison experiences can strengthen the sense of identification with a criminal caste and confirm felons in their life course rather than rehabilitating them. The British practice of exporting their felons to distant colonies is an interesting natural experiment in rehabilitation because dangerous criminals were released in large numbers in the Australian and New Zealand colonies. Did they continue to express their criminal inclinations? Or did they buckle down to the hard work and responsibility of life in a new colony?

The answer is not nearly so clear-cut as we would like because such colonies typically have a large proportion of men in the population, which inevitably translates into high crime rates. Yet, there are at least two reasons for optimism that criminals may be rehabilitated by having an opportunity to participate in the community as a free person. The first is that although Australian crime rates were high, they were not higher than the crime level in America's Western frontier that also flourished in the middle of the nineteenth century. The second reason for optimism, as noted earlier, is that Australian crime rates fell steadily as the colony became more civilized and attracted female settlers in larger numbers. Whether the first generation of transported felons were rehabilitated or not, their descendants were not particularly likely to be criminals contrary to the genetic determinist hypothesis. This conclusion is similar to the case of Pitcairn Island, a much smaller community founded, perhaps entirely, by "criminal" males and their Polynesian mates.

Colonies are not a good place for rehabilitation, as everyone who has ever watched a Western movie is aware, fictionalized, exag-

Fig. 6. John Dillinger, a celebrity criminal. (Library of Congress)

gerated, and romanticized though these depictions are. America's western mining towns were some of the most violent communities inhabited by Europeans at the time, even though they were not set up as places to siphon off dangerous criminals. Nevada County in California was one of the most violent such communities. During the late 1850s, the homicide rate there was twenty times that of contemporary Boston.[18] Fort Griffin, Texas, a haunt of soldiers and buffalo hunters, was even more violent with a homicide rate almost forty times as high as that of Boston. Such communities had very high crime rates for several reasons, including their attraction for young adventurous men, abuse of alcohol, availability of guns, and lack of adequate policing. Crime frequently paid well and there were often few consequences for acts of appalling savagery and selfishness. The New South Wales colony was similar. It was a defector's paradise, exactly the wrong sort of community into which criminals ought to be released if they are to be reformed.

In recent years, prison administrators have been emboldened to experiment with open prisons in which serious but currently nonviolent prisoners can be gradually reintegrated into the mainstream community. Some of the most interesting such experiments have involved gardening. Afficionados of gardening believe that there is something intrinsically healing to the criminal psyche about caring for plants. Whether this is true or not, involvement of prisoners in the cultivation of vegetables and flowers has emerged as a useful bridge that helps them to adjust more smoothly from the challenges of prison existence to the very different challenges of establishing an independent life following the end of a long sentence.

REHABILITATION BY GARDENING

The case for gardening is made in a direct and fairly compelling manner by the English movie *Greenfingers*, which chronicles the true story of a group of inmates at Leyhill, an open prison in England, who realized their ambition of competing at the Chelsea Flower Show. Although unsuccessful on their first attempt, the prison gardeners ultimately won the gold medal at Chelsea, which is Britain's highest gardening honor. They wowed judges and the general public with their incredibly detailed gardens of native species representing themes of regeneration and redemption.[19]

There are perhaps many different reasons why gardening projects were so helpful in the rehabilitation of Leyhill's prisoners. Gardening provided an escape from the dreary routine of prison life with prisoners standing outside their cells five times per day to be counted, and all of the other numbing routines that inmates endure. Those on gardening duty could spend their entire day in the open air, becoming absorbed in their projects and essentially forgetting that they were imprisoned felons. Gardening also builds patience and discipline because each step requires time before one sees any progress. Cooperating in the grand scheme of producing a prize-winning garden at Chelsea creates a collaboration among prisoners for good ends. Left to their own devices, they might have been planning a grandiose bank robbery together. There is clearly altruism among prisoners that is normally directed toward selfish (defector) objectives but can be redirected to shared projects that benefit the larger community.

Whatever about the success story of the handful of talented individuals who put together Leyhill's prize-winning exhibit at the Chelsea garden show, it is clear that working in gardens can be beneficial for the rehabilitation prospects of larger numbers of prisoners. Over a hundred of Leyhill's inmates work in the glasshouses, arboretum, museum, and garden shop, generating annual revenues of about $4 million. Such enterprises bring prisoners in touch with the mainstream community and provide them with opportunities to adjust to regular employment.

In San Fransisco's prisons, gardening has been used quite deliberately as a tool for reintegrating released offenders into the community. Many long-term prisoners find life outside jail extremely challenging because of the problems of finding employment and obtaining a safe place to sleep. When San Francisco's prisoners were released, some found employment in the Garden Project, a fruit-and-vegetable enterprise spawned by the jail's farm.

The Garden Project is the brainchild of Catherine Sneed, a counselor for women prisoners who became inspired by John Steinbeck's message in *The Grapes of Wrath*,[20] that working in the soil heals the human psyche. She began with a prison garden. Produce from the garden supplies charities like food kitchens and AIDS pantries. After local business owners observed the prison garden, they suggested that vacant lots might be used to grow supplies for local restaurants. The Garden Project is located outside the jail and its produce can therefore be sold.

By 1997, over one hundred prisoners were volunteering for garden work each day. Many of these could graduate to the Garden Project at the end of their sentence. This gradual transition to life outside prison facilitated rehabilitation. Four months after release, only 6 percent of the gardeners had committed another offense, as compared to 29 percent of the other ex-prisoners.[21]

As always, in the case of such natural experiments, it is difficult to know why gardeners did better. They might have been less serious offenders to begin with. They might have been healed by contact with the soil, as Steinbeck and some prison administrators believe, or they might just have benefited from rapid recruitment to paid employment. The beneficial results are also consistent with the idea that most people want to cooperate when given a reasonable opportunity to do so. Reciprocity is built into all of us and craves expression. In other words, we want to be a part of something that is larger than our immediate selfish interests. As we will see, reciprocity not only makes us happy but actually contributes to our health from very early in life until old age.

PART 3

The Social Impact of Kindness

CHAPTER 6

Kindness and Health

A capacity for altruism is built into human beings by natural selection. This assertion is supported not just by what we do but by what goes on underneath the skin. Converging lines of evidence show that social support can have profound effects on health and length of life. Even living without one's father during childhood can undermine happiness and health. Being able to help others also contributes to contentment and well being. These fundamental aspects of human sociality are the product of a long history of evolutionary design going back to the earliest mammals.

THE VOLE'S STORY

According to French microbiologist François Jacob of the Pasteur Institute and College de France, evolution can be thought of as a tinkerer who adapts whatever materials happen to be available rather than making them from scratch, as an engineer would do.[1] Thus the fine bones of the middle ear in mammals, the hammer, anvil, and stirrup (*malleus, incus,* and *stapes*) are derived from jaw bones of reptiles that were co-opted for their new role in hearing. Evolution is conservative: it keeps old stuff around. This applies as much to social behavior as it does to anatomy. That is why such an unlikely animal as the vole, or field mouse, can tell us a great deal about social attachment in humans.

Different species of voles look alike, but they have remarkably clear differences in their social behavior. Prairie voles (*Microtus orchogaster*) are a great deal more sociable than those living in mountainous regions—mountain voles (*Microtus montanus*)—which are solitary. Adult males and females meet only in the context of mating, and females mate with several males. Even the bond between females and young is weak: females abandon the pups about two weeks after birth, and young show little distress at the separation. By contrast, prairie voles could be described as having family values. They are monogamous with a mated pair staying together during the breeding season and cooperating in the care of their young.[2]

The very great difference in social behavior between these two closely related species has excited a great deal of research interest because it allows scientists to investigate the biological underpinnings of social behavior. Most of this effort has focused on differences in the hormone oxytocin, the "cuddling hormone," that is released into the bloodstream from the pituitary gland at the base of the brain and interacts with receptors throughout the body, and the brain itself. Hormones affect behavior as a result of their interaction with specialized receptors in the brain, and researchers map the oxytocin receptors at various sites in the brain in search of a biological explanation for the differences in social behavior between the two species of vole. The oxytocin receptor is a large protein molecule to which the neurotransmitter attaches, thereby allowing communication between brain cells.

One important finding, by Thomas Insel of the Center for Behavioral Neuroscience at Emory University and his colleagues, is that mountain voles have fewer receptors for oxytocin than their prairie cousins in key areas of the brain that control social and sexual behavior, particularly the hypothalamus and amygdala.[3] Not content with demonstrating such correlations, researchers have conducted experiments showing clearly that a monogamous female's tendency to stay with her mate following mating is due to the effects of oxytocin. The hormone is released in large quantities during copulation and bonds the female to her mate. If oxytocin receptors are blocked, however, a pair bond does not form. The hormone vasopressin is also critical for pair bonding in males. Both oxytocin and vasopressin are produced in the hypothalamus of the brain and released to the bloodstream via the pituitary gland.

Oxytocin receptors are differently distributed in the brains of the two species of voles, which helps us to understand the differences in their social behavior. Prairie voles have more oxytocin receptors in parts of the brain that mediate pleasure, specifically those that are implicated in addictive drugs. When monogamous voles mate, it is as though they received a shot of an euphoriant drug that addicts them to their partner.[4]

Oxytocin and vasopressin systems not only account for monogamous mating in prairie voles, but are also responsible for parental behavior in males. Hence they bolster the argument that these neurotransmitters can account for the entire suite of family values in males as well as females.

One ambitious experiment used viruses as a vehicle to carry the gene for the vasopressin receptor into specific sites in the vole's brain. In this way, the number of receptors was increased. According to one of Insel's collaborators at Emory, Larry Young, voles whose brains were altered in this way were more social. They formed pair bonds even without copulating. It was love at first sight, according to Young.[5] It was certainly reminiscent of human romantic behavior in sexually restricted societies wherein people fall in love and form a pair bond in advance of marriage and sexual intercourse.

OXYTOCIN: THE CUDDLING HORMONE?

To those who are unfamiliar with neuroscience, it may be surprising that such large differences in kindness between prairie voles and mountain voles could be the product of alteration in just one or two neurotransmitter systems of the brain. The implications for humans are potentially enormous. We can look ahead to a future where it is possible to treat children suffering from autism, which is essentially a disorder involving underdeveloped sociability. What about the potential for turning selfish individuals, such as criminal sociopaths, into model citizens? Such projects are clearly science fiction at this point.

Leaving aside the ethical controversies, research on the chemical basis of sociability in voles is theoretically very important for students of altruism. One reason is that it unites various types of social cooperation among mammals, allowing us to see them as the product of a generalized increase in oxytocin/angiotensin-mediated sociability.

The biological basis of altruism among mammals, including humans, is clearly different from that for insects because mammals have more flexible behavioral control systems orchestrated by emotional reactions like fear, anger, and social attraction. Insects seem to be much more robotlike in their social behavior. Their actions are comparatively simple, rigid, and relatively independent of varied sensory feedback while they are being carried out. If a bee chases after us with a loud buzzing sound as it attempts to sting us, we might imagine that the animal was angry, but current knowledge of the behavioral control systems of insects suggests that they do not have emotional reactions analogous to those of mammals. They certainly do not have anything resembling the brain structures regulating mammalian emotions.

In mammals, altruistic actions are supported by social emotions, by kindness toward others. Hence the title of this book. Mountain voles are fascinating precisely because of their lack of kindness. There is little warmth even in the relationship between mothers and offspring. When mothers are removed from the young in experiments, there is none of the separation distress that one sees in most other animals. If young are removed from a mother, she doesn't even bother to look for them.

Young mountain voles have little interest in peers as social companions, treating them entirely as competitors. In the terminology of the prisoner's dilemma game, mountain voles are defectors. If they catch sight of another vole in the vicinity of their burrow, they will generally chase the intruder away, thereby defending precious food resources.

In contrast, prairie voles are cooperators, generally avoiding aggression against neighbors. Why this difference evolved is not entirely clear, but one can speculate that the sparse food resources available on mountains forces voles to disperse with solitary individuals migrating to patches having the best food resources and refuge from predators. For prairie voles, the comparatively even distribution of large amounts of food permits much higher population density.

It is even possible that being close to others of the species provides an advantage in terms of defense against predators. An individual is less likely to be captured by a predator if there are thousands of other potential prey that can be eaten first. Similar reasoning has been applied to formation of large herds of grazing animals, like wildebeest, in Africa and to schooling in fish. This idea was first developed by English biologist William Hamilton, who coined the term "selfish herd."[6]

Whatever the evolutionary origin of social differences in voles, these differences help us to see a wide variety of altruistic social behaviors as going together, from care of the young to social toleration of other members of the species to the social support between parents for the purposes of raising young, a form of altruism that is rarely described as such, perhaps because it seems so common and because its interpretation is so obvious.

As far as our own species is concerned, there is reasonably good evidence that our sociability is affected by oxytocin too. Oxytocin dulls pain, promotes peace and relaxation, and has an anti-inflammatory action that protects the body against chronic illnesses such as heart disease. It plays an important role in the pleasure that people may feel in each other's company, particularly in intimate relationships.

Oxytocin is released in the bloodstream of nursing mothers for whom it not only plays a critical role in the milk letdown response, but also reduces any pain and irritation of the nipples caused by breast feeding. It also promotes calmness and social bonding. Some mothers report that nursing is pleasurable, and it is probably no accident that oxytocin is released in large quantities during sexual intercourse. It evidently plays a role in promoting orgasm for both sexes.[7]

Recent evidence indicates that oxytocin levels of the blood increase in the context of all types of affectionate relationships. The connection between oxytocin and close bodily contact has led to its being described as the cuddling hormone. It may seem that there is a contradiction between a hormone promoting the intense sexual pleasure of orgasm and being implicated in the milder pleasures of less intimate social interaction. This is easily explained in terms of amount, however. Thus, different opiate drugs may either produce intense pleasure, such as that of a heroin user, or merely dull pain, as in the case of opiate painkillers. The distinction is largely a matter of dosage; opiate addicts who cannot obtain their drug of choice may switch to using large quantities of pain medication.

Whether the dosage is high or low, opiate drugs are all highly addictive. This can have tragic consequences for people who rely on artificial sources for their high, and on all those who are touched by the problems of chemical addiction. Oxytocin is also "addictive" in another sense: it promotes long-term social relationships because the individuals obtain pleasure in each other's company.

Everyone is conscious of the role played by sexual pleasure in binding people together, particularly early in a sexual relationship, which is an exact analogy to the pair bonding through sexual pleasure of voles. Yet, the intense rush of sexual pleasure is not enough to keep people together. They must be compatible. They must make each other feel good on a continuing basis. Merely having a sexual relationship with another person does not ensure the formation of a pair bond. In fact, most human sexual relationships, in a sexually liberated society such as our own, are transitory.

Sexual pleasure undoubtedly plays a role in bonding people together, but it is not an essential feature of a stable marriage. In some societies, such as the Yoruba of Nigeria, there is a postpartum taboo on sexual intercourse that may last for a year, or even two, following the birth of a child. The function evidently is to space out births. Societies with postpartum taboos do not have high levels of marital instability associated with sexual abstinence. Moreover, there is a steady decline in frequency of sexual intercourse with the duration of a marriage, but this does not undermine marital stability. As the sexual excitement in a marriage declines, there is often an increase in a couple's affection and sense of companionship.[8]

The pitch of romantic excitement declines, but it would be a mistake to underestimate the strength of a pair bond between a couple who have lived together for decades. When they are separated for only a few days, they may miss each other intensely and become irritable and restless, like opium addicts experiencing withdrawal symptoms. The loss of a loving spouse is one of the severest emotional losses in a person's life, and it can precipitate a severe depression.

Oxytocin is not the only chemical acting on the body and brain to promote social bonds, of course. Its effects are augmented by opiumlike neurotransmitters known as the endorphins, or natural opiates. Some of these substances are released by the pituitary (as oxcytocin is) and act on the brain as well as the body. Endorphins are released in response to painful stimulation, such as a bruise or cut, and damp down pain sensations to more manageable levels. They are also released in response to psychological stressors. Thus, a rat experiences temporary loss of sensitivity to pain when it is in the presence of a cat, even though the cat cannot reach it.

Endorphins are released also in response to skin contact such as that of caressing an infant or hugging a child. Like oxytocin, they are also released in response to stimulation of the genitals during sexual intercourse.

By emphasizing the chemical basis of affectionate relationships, there is no desire to imply simplistic chemical reduction. Even in rats, there is an important emotional component to these chemical mechanisms. The point, instead, is to emphasize the evolutionary continuity of affectionate relationships. Social attachments are mediated by much the same chemistry in human beings as in other mammals. This is not such a large claim as it might seem when you realize that closely related animals, such as different species of voles, can respond very differently to the same kind of social event. Even though the hormones circulating in their bloodstream are much the same, the way in which their brains respond is very different because of variation in brain receptors.[9]

The chemical basis of social bonds is important to the study of altruism for the obvious reason that we are more likely to help people if we have chemical bonds of affection for them. This explains a great deal of the altruism within families, whether the people are genetically related, like parents and children, or unrelated, like husbands and wives. It also accounts for reciprocal altruism among close friends. These mechanisms are not specific to humans, of course, but are a part of our mammalian heritage.

Most close relationships are mediated by physical closeness in humans as well as other mammals. Nonsexual friendships among adults can be maintained by social interactions involving very little physical contact, of course, but as you go backward into childhood, the modality of touch becomes increasingly meaningful. Touch is the first indication that a baby has of its relationships with the mother and other adults. Developmental psychologists discovered that a physically close relationship with adults early in life can be essential for normal brain development and health. It also plays a role in socialization and the emergence of altruistic tendencies.

EARLY TOUCH AND HEALTH

Interest in the role of touch in infant development was sparked by research showing that premature infants do better if they are massaged. Preterm infants who were massaged for forty-five minutes per day (in three fifteen-minute sessions) for ten days, as part of an experiment conducted at the University of Miami Medical School, gained 47 percent more weight than premature infants who were not massaged. The treated infants also scored higher on the Brazelton test of behavioral responsiveness. Their improved condition meant that they could be released from a hospital six days earlier than children who were not massaged. Interestingly, the infants receiving massage therapy were still heavier at the age of one year, as well as scoring higher on motor and cognitive development. This fascinating result has been repeated by researchers working in various countries, including Israel and the Philippines.[10]

Why did the massaged preemies gain more weight? The most obvious answer, that they ate more, turns out to be incorrect. Neither did they spend more time asleep, which would have minimized energy expenditure. A possible explanation is that the massage increased stimulation of the vagus nerve, which enhances the release of insulin, the hormone that draws nutrients from the blood into tissues.

Infants normally receive a great deal of touch stimulation early in life, so that the research on ways of using touch to promote health and brain development in premature infants is also being examined in relation to the effects of tactile deprivation for infants kept alive in incubators.

Increasing touch stimulation for full-term infants can have a variety of beneficial effects, too. Infants who are massaged in the first three months of life are both more alert when awake and calmer, as determined by measurement of the stress hormone, cortisol, taken from saliva samples. Like the preemies, they gain more weight.

Provocative as these results are from the point of view of promoting infant health and well being, the effects on social development are even more fascinating. Infants receiving massage scored higher on emotional expressiveness, sociability, and capacity to be soothed by adults when upset. In short, they were more socially developed and less anxious and depressed. Reduced anxiety could be explained in terms of their hormone profile (reduced cortisol, norepinephrine, and epinephrine). Infants who were massaged were more active and more animated. These phenomena reflected a high level of brain serotonin that acts as a natural antidepressant.[11]

Many of the effects of therapeutic massage are due to increased relaxation responses and reduced secretion of stress hormones. One way of thinking about these phenomena is in terms of the feeding patterns of mammals. When they are relaxed, they feed and digest their food. When alerted to the presence of a possible predator or other enemy, a fight-or-flight response is evoked that increases physiological arousal through the release of stress hormones. Not only do they stop feeding to cope with the threat, but their digestion shuts down so that resources, such as oxygen and sugar and the blood that carries them, can be diverted to skeletal muscles that orchestrate the fight-or-flight response.

The body's autonomic nervous system has two branches, the sympathetic nervous system that organizes fight and flight and the parasympathetic system that controls digestion. When a person is anxious, the parasympathetic system is shut down so that a common sign is dryness of the mouth. Much of the effects of massage therapy are consistent with stimulation of the parasympathetic nervous system, which accounts for weight gain of premature infants as well as their calm alertness.

When the relaxation-promoting parasympathetic system is in charge, stress is reduced. This has two key implications for infants and children. One is that copious and continuous release of stress hormones undermines the immune system. Conversely, tactile stimulation would thus promote immune function. The other major implication is that parasympathetic activity produces calm alertness that is good for mental activity. That is why massaged infants do better on measures of cognitive development. Their brains function better because they are less anxious. Anxiety over prolonged periods may also cause clinical depression, one of whose symptoms is slowing down of thought processes.

Clearly, it would be a mistake to exaggerate the role of tactile stimulation of infants to the exclusion of other sensory modalities. Yet, young children are very much oriented toward physical interaction with the people and objects in their environment. Tactile stimulation goes along with other forms of social interaction, particularly speech, which may matter even before birth and becomes increasingly important with the child's development of language competence. That said, the ground rules of social relationships are laid down when infants are preverbal (i.e., cannot yet express themselves in words, although they respond to the words of others around them). People first learn to be social through touch, and these early experiences affect future social relationships. Physical closeness early in life not only affects the emotional closeness of future social relationships, but also affects altruistic tendencies in adult life.

Parents who are close to their children, in an emotional sense, are also physically close to them and engage in a great deal of bodily contact. In many subsistence societies, infants spend most of the first year of life in close proximity to the mother, sleeping next to her at night and being carried for much of the day. Cross-cultural research finds that children in violent societies are relatively touch-deprived. For example, among the relatively peaceful Arapesh of New Guinea, mothers carry their infants in net bags so that they are held next to the mother's skin. In contrast, Mundugumor infants also of New Guinea are carried in baskets, out of contact with the mother. The Mundugumor are an exceptionally violent society with a reputation even for cannibalism.[12]

Such research excites a fair amount of skepticism, in part because it is largely irrelevant to war-making of state-level societies. Thus, the Japanese, who have a comparatively high level of physical contact, have a history of foreign aggression and occupation. Americans, who have a relatively low level of physical contact, have engaged in relatively little foreign invasion, historically speaking, but have taken part in many wars at any rate. Deprivation of contact can promote aggression at an individual level, however. In fact, there is good evidence that many of the most violent criminals in our society have been deprived of maternal affection.[13] The difference is that the level of hostility among individuals is predictive of war in simple societies where each individual has some decision-making input. In large complex states, individual soldiers of the rank and file have little or no impact on decisions about going to war that are generally in the hands of politicians.

We are inclined to see aggression as pathological, but a child who is carried around in a basket may be just as healthy physically and psychologically as one who is carried around in a net. They are just more likely to defend their interests using violence. From an evolutionary perspective, the amount of gentleness experienced by young children is predictive of their altruism toward others. Conversely, a physically and psychologically harsher early environment toughens people.

In addition to minimizing physical contact between mothers and infants, warlike societies often emphasize corporal punishment in disciplining children. Many different lines of evidence suggest that being beaten makes children more aggressive. This is just as true of family research in the United States as it is in comparisons between different societies studied by anthropologists.[14] An evolutionary interpretation holds that aggressiveness is an adaptive response to being raised in a harsh and violent society.

The Spartans, one of the most violent societies in history, exploited this developmental process to raise warriors. In order to toughen up their boys, they deprived them of sustenance, forcing the youngsters to steal food. Spartan military thinkers believed that this would hone skills at creeping up on an enemy undetected. Whatever about stealth, there is little doubt that facing such challenges early in life forces people to be tough and to look out for themselves.[15]

The real pathologies from social deprivation are seen in children unfortunate enough to be raised in orphanages that used to be run either like prisons, or like hospitals. In many hospital-type institutions, children were deprived of virtually all social contact in the interests of controlling infectious diseases. Such children appear to do reasonably well in the first half-year of life, but there is a marked decline in measures of infant intelligence thereafter, as children are deprived of the social stimulation necessary for learning to speak and surmounting the other developmental hurdles crossed by children receiving normal stimulation from their mothers. Such declines in infant intelligence due to social deprivation are largely reversible when children are adopted out of orphanages into families in the first few years of life, however.[16]

The importance of tactile stimulation for social development of primates was highlighted in Harry Harlow's well-known research on the development of attachment to mothers in rhesus monkey infants. The goal of this work was to separate those aspects of mothers needed for infant-mother attachments to form.

TOUCH AND MONKEY LOVE

In the wild, infant monkeys spend much of their time in close contact with the mother just as human infants do in subsistence societies. When the mother moves over long distances, the infant rides on her back, clinging tightly to the fur. Interestingly, newborn human infants have a very strong grasp reflex that might be an evolutionary vestige of a mechanism for clinging onto mothers. Harry Harlow wondered which aspects of the infant's experience allowed it to become bonded with the mother.[17]

Harlow's best-known experiment provided mother-deprived infants with two kinds of artificial mother. One was a wire mother from which a feeding nipple protruded. The infant monkey obtained its nourishment there. The other artificial mother was like the wire mother but she was covered in terry cloth and did not provide food. Which dummy mother would the infant prefer, the one who provided food, or the one who was soft and cuddly? The result was clear, monkeys visited the wire mother only when they needed food. They preferred to spend time in contact with the cloth-covered mother. What is more, they derived emotional support from the cloth-covered figure. If anything frightened them, they rushed to the cloth mother and clung on tightly. These responses are remarkably similar to the behavior of young children who hug their stuffed toys and will often cling to them when distressed. Similarly, many children have a soft blanket that they use to seek comfort in similar situations.

The bottom line from Harlow's work is that primate emotional attachment to mothers is partly based on physical contact and touch stimulation. The sense of touch is very important for social development in human children too. One of the most significant types of deprivation experienced by orphanage children is lack of physical contact from caring adults. Such lack of social and emotional contact apparently impedes the formation of intimate relationships in later life. Children raised in orphanages have great difficulty in forming close friendships and have trouble getting close to others in sexual relationships.[18]

Deprivation of close physical contact with others affects health and growth as well as social development. After the fall of Communism in Romania, people around the world were shocked at the appalling conditions inside Romanian orphanages. As well as showing signs of intellectual backwardness, these unfortunate children were very small for their years. According to Tiffany Field, a leading researcher in the field of therapeutic touch and director of the Touch Research Institute at the University of Miami School of Medicine, massage therapy was used to stimulate physical growth of Romanian orphans. Early physical contact is also important for developing social trust, which is a vital component of altruism.[19]

PHYSICAL CONTACT AND HUMAN TRUST

Physical contact with a caregiver is the primal scenario in which a child learns to trust other people. To some extent, close attachment to mothers is a biological given. To some extent it must be earned by the mother through sensitive responsiveness to the needs of the infant. A sizable proportion of infants, about one-third, do not have secure emotional attachment to their mothers. The problem is particularly acute in the case of irritable infants who seem to fuss no matter what the mother does to comfort them. Even in these difficult cases, the mother's behavior is of critical importance. Mothers often respond to the constant fussiness by emotional disengagement, which aggravates the problem. When they are trained to interpret the infant's nonverbal signals correctly, the relationship between child and mother typically improves. For example, if a child is overstimulated, it turns its face away, and at that point the mother should stop talking to it. When it is ready to interact, it will indicate this by looking toward the mother again.[20]

From a developmental perspective, it is not really possible to separate physical closeness between children and adults from the dimension of emotional trust. This is of considerable importance in the development of altruism because altruistic sentiments and actions are founded on a basic trust of other people. If this fundamental connection is accepted, then the correlation between physical closeness of children to parents and their altruistic tendencies makes sense. This connection has been established in the research of Tiffany Field and her colleagues.[21]

In one project, Field asked high-school seniors a great many questions concerning their relationships with parents, siblings, and peers. She found that students who were emotionally close to their parents had a close physical relationship involving frequent bodily contact, such as hugging. Moreover, children who were close to their parents also had better relationships with other family members and with peers. They were clearly more altruistic in their outlook, scoring lower on aggression and higher on sympathy for other people. These conclusions are remarkably consistent with the finding that infants receiving massage therapy are more socially developed by the age of two years.[22]

Trust, based on physical closeness, begins early and forms the foundation for subsequent trusting relationships. Developmental psychologists found that a secure early relationship with parents is critical for many future kinds of social interactions requiring trust. Emotionally secure children are more intellectually curious, have less fear of failure, and work harder at school, and ultimately, at careers. They have more successful long-term relationships with romantic partners and spouses. They are even more involved in their jobs and careers.[23]

The connection between physical intimacy and altruism is not just a vestige of childhood and the bond of intimacy between mothers, fathers, and their children. It can be observed among adolescents and adults. Researchers in the field of spatial behavior have long been aware that people in different societies use space differently. Friends in Sweden keep each other at longer distances than friends in Italy, for example. Presumably, friendship is experienced differently in these two countries and varies along dimensions of intimacy and self-disclosure. It would be a mistake to draw any more general inference about the degree of altruism of people in one country compared to the other, of course. Yet, it would be reasonable to assume that altruism is directed preferentially toward those individuals with whom we are physically close, specifically lovers and children.

Tiffany Field conducted a study of touching among adolescents in France compared to the United States as observed at McDonald's restaurants.[24] France is a high-contact society, and French adolescents spent a lot of their time (43 percent) expressing affection for peers through the medium of touch (leaning against them, stroking, kissing, and hugging). American adolescents were far less demonstrative, spending only 11 percent of their time engaged in affectionate contact with friends sitting next to them in the fast-food restaurant. Interestingly, American adolescents spent far more time touching themselves, preening their hair and clasping their hands. American adolescents were also more aggressive in their conversation and gestures, prompting Field to speculate that the greater degree of violence in the United States compared to France might reflect deprivation of touch. In support of this claim, she invoked Harlow's work on touch-deprived monkeys that grow up to be more hostile toward peers.

Among their other neurochemical abnormalities, touch-deprived (i.e., mother-deprived) monkeys have depleted serotonin (and norepinephrine) levels in their brains, which means that they have trouble controlling their hostile impulses. Extremely aggressive children have a similar neurochemical profile that can result from early deprivation of close social contact.

Deprivation of physical closeness is not the complete picture, of course. Orphanage-raised children are often touch-deprived but are rarely very aggressive. They more often seem compliant and lacking in emotional depth.[25]

Physical closeness, and the emotional closeness of which it is the expression, are essential for the development of empathetic and altruistic dispositions toward others.[26] It might be imagined that children who grow up relatively lacking in empathy would at least look out for number one. Health professionals find that the opposite is true. In other words, young people who grow up in a harsh social environment not only develop a tough attitude toward others, they also learn to be hard on themselves. This phenomenon emerges in a number of different contexts, including risk-taking behavior, drug abuse, and failure to take care of one's health. One criterion of a harsh early environment is absence of the father from a child's home. Children raised without their fathers live in a less healthy manner and experience poorer health throughout their lives, on average. Father absence can also produce a decline in altruism, and can even manifest itself in an increased risk of becoming a criminal offender.[27]

FATHERLESS CHILDREN: THE HEALTH IMPACT

Much has been written about the impact of fathers on their children, an issue that has come to the fore in recent decades of marital instability. It is still quite difficult to work out why children raised without fathers have poorer social and economic prospects and are at greater risk for a multitude of health problems. Psychologists point to deprivation of paternal interaction and to the feelings of sadness and abandonment of mothers who feel let down by the fathers of their children and communicate these feelings at some level to the children themselves. Sociologists emphasize the economic impact of single parenthood. Children raised in a single-income family are more likely to be poor and to live in a dangerous, crime-ridden neighborhood. Biologists, medical researchers, biological psychologists, and biological anthropologists focus on the way in which stressful environments alter childhood stress hormones, and thereby undermine immune function. Though all these approaches have something going for them, it is too soon to claim that the health problems of father-absent children have been adequately explained in terms of cause and effect.

Even where families are not wretchedly poor, children may suffer from absence of their fathers. This can be illustrated by middle-class American families in which the parents are split up by divorce. Problems experienced by children whose fathers are absent due to parental divorce included the following:[28]

- increased risk of school dropout (by 150 percent for white children, 100 percent for Hispanics, and 76 percent for blacks).
- increased likelihood of divorce themselves (60 percent more likely for white women and 35 percent higher for white men).
- increased risk of teen pregnancy (by 50 percent).
- increased incidence of social, emotional, and behavioral problems and learning difficulties (by 200–300 percent).
- increased crime rates (by about 50 percent).

Each of these outcomes can be interpreted as a health issue because there are many overt links to poorer health. Thus, high-school dropouts are more likely to work in undesirable jobs with exposure to occupational dangers and unpleasant work conditions that might include greater exposure to noise, repetitive stress injuries, and pollution. Similarly, poor people are at risk of unhealthy sedentary lifestyles and the eating of junk foods that undermine health. Apart from such environmental connections between parental divorce outcomes and health problems, there is the physiological effect mediated by psychological stress.

In some ways, evidence establishing the importance of a father's presence in the home has surprised psychologists. The history of psychology is infused with the assumption that only the mother matters for child development. Recent evidence places this assumption in doubt. Thus, young children are often strongly attached to their fathers as well as their mothers, and some have a preference for fathers as companions and playmates, even though they may retreat to the mothers' arms when distressed or hurt.

Fathers play more vigorously with children, something that alarms many mothers. This difference in interaction style can have important consequences. Among premature infants, those whose fathers played with them often had superior cognitive development compared to those whose fathers were less interactive.[29] This finding makes sense in view of Tiffany Field's research showing that massage therapy improves brain development in children. Both vigorous, playful stimulation and massage therapy each involve stimulation of deep pressure receptors in the skin.

Characteristics of fathers, such as level of education, income, and residence with children, may also be as important as characteristics of mothers in predicting children's educational achievement and avoidance of delinquency. How could fathers have such an important effect given that many spend little time with children? The answer may lie in this very variability. Most children are loved and protected by their mothers. Most mothers give everything that they can, emotionally and economically, to their children. Maternal love and support is pretty much a constant, but paternal love and support is variable. Children respond strongly to this variability as can be seen in the differences between children of single mothers compared to two-parent families.[30]

The causal path from father absence to social problems is imperfectly understood. One of the most intriguing explanations to emerge from recent research is that homes that have absent fathers are psychologically stressful ones. Chronic stress in childhood leads to emotional problems, such as depression and anxiety. A stressful early life may predispose children to poor academic performance, delinquency, drug abuse, and other social problems as well. Whatever about the exact sequence of causes and effects, there is no doubt that homes in which fathers are absent are psychologically stressful for children.

This conclusion emerges from Mark Flinn's study of Dominican children.[31] Flinn, an anthropologist at the University of Missouri in Columbia, gave children in his study villages chewing gum that they spit into a sample container. In this way, he obtained a saliva sample that could be used to measure levels of stress hormones, particularly cortisol. As he walked about the study's villages, he also collected information about changes in family structure, such as parental separation, and about children's health. His detailed data allowed for a rich matching up of family events and children's hormonal response. The main conclusions were that being raised without a father is stressful, as manifested in elevated stress hormones, and that children living stressful lives had weaker immune function, as reflected in more frequent colds and absences from school due to various ailments.

It is unlikely that Flinn's findings are peculiar to Dominicans so that absence of a father may be stressful in other societies too. Why is absence of their fathers so psychologically stressful for children? Science has not yet provided clear answers, but there are many possibilities:

- Economics may be important: it could be tough for mothers to provide for their children even if they have the help of extended families.
- Mothers may feel anxious and depressed as a consequence of raising children alone and transmit these feelings to children.
- Children may feel sad and abandoned if they no longer see their fathers on a regular basis.
- Following departure of biological fathers from the home, the mother may have live-in lovers, which can be a focus of discord as the lover and the children compete for the affections of the mother.
- Families without adult men may feel more anxious as to their physical safety.

The issue of physical safety may seem less important than the others, but it is not hard to imagine some environments, such as dangerous city slums, in which the presence of a child's father would be somewhat reassuring. In subsistence societies, presence of the father is critical for the safety and survival of children. Among the Ache hunter-gatherers of Paraguay, for example, survival is tough at the best of times. About one-fifth of children living with both parents die before the age of fifteen years. If their father leaves or is killed, twice as many children die (45 percent). The increased mortality is due to greater homicide victimization, as well as greater vulnerability to disease, which could reflect both poorer nutrition and suppression of the immune system due to psychological stress.[32]

A stressful childhood environment obviously has deleterious effects for health because stress hormones weaken immune defenses against illness, potentially setting children up for a lifetime of poorer health. The absence of a father is not only associated with poorer childhood health but also affects health throughout life, partly because habits of anxiety and depression established in childhood continue to affect health, and partly because of poor health habits.

Between the effects of stress and unhealthy patterns of adult behavior, such as drug abuse, it is hardly surprising that children raised without their fathers live shorter lives, even when they do not lack material comforts. This phenomenon was noticed in a study of longevity among gifted children. Men whose parents divorced before they reached maturity were three times as likely to die by the age of forty years.[33] For both sexes, parental divorce subtracted an average of four years from the lifespan. Interestingly, the early death of a parent did not have such marked effects on longevity, suggesting that the phenomenon was due to the social stress associated with divorce. Such early psychological stressors can undermine health later in life. They also affect health behavior.

HEALTH BEHAVIOR AND ALTRUISM

A psychologically stressful early environment undermines altruism, as already discussed. When a child is confronted by a family environment in which other people behave in selfish and deliberately irritating ways, there is little point in adopting an altruistic policy because she will only get taken advantage of. In a community of defectors, cooperation cannot work. The Hart and Ridley study of language development provided unmistakable evidence that children mirror the level of altruism of parents in their own conversations. What does this have to do with health? The answer is that children's attitudes to themselves, to their own body, and to protecting it are very much a function of their level of altruism toward others.

It is philosophically fascinating that selfish people should be just as incapable of showing concern for their own health and long-term welfare as they are of being concerned about others. Difficult early environments *do* predispose people to looking out for their *immediate* interests (i.e., comforts and pleasure). That is one reason why being raised in poverty is not only a risk factor for criminal behavior but also a risk factor for illnesses brought on by self-indulgence of various kinds, including drug addiction, alcoholism, venereal diseases, and the cardiovascular problems attributable to junk food and a sedentary lifestyle, which are strongly associated with poverty in our society.

The connection between venereal diseases, including AIDS, and family background is easily made. Children who have distant or conflicting family relationships are likely to be sexually active from an early age, to have unprotected sex, and to be active with more partners. Early sexual activity is a function of lax parental supervision and is more likely for children who live in neighborhoods inhabited by delinquent teenagers.

How do we explain the paradox of selfish people behaving in ways that are likely to injure their health? People who do not see themselves as connected in meaningful ways to larger entities, whether families, local communities, church congregations, political parties, sports clubs, or work colleagues, cannot commit themselves to activities on behalf of others and are thus largely confined to the pursuit of egoistic goals—pleasure. In the language of evolutionary biology, they are competing for the resources necessary for survival and pursue short-term reproductive tactics, that is, sexual intercourse without emotional or financial commitment.

Maximizing pleasure does not always work out well in the modern environment with dangerously addictive drugs, prevalent venereal diseases, and such an abundance of food, at least in affluent countries, that overeating is a real health threat. Thus dietary obesity is a major cause of leading killers like heart disease, hypertension, strokes, and diabetes.

From this broad evolutionary perspective, it is interesting that there is a conspicuous difference between men and women in health behaviors and risk-taking. Evolutionists explain the poorer health behavior of men in terms of an evolutionary history of mating competition that called for opportunistic exploitation of casual sex and a corresponding willingness to contemplate aggression from reproductive rivals. Natural selection has arguably pushed women in the opposite direction so that they take better care of their bodies and health.

SEX DIFFERENCES IN HEALTH AND NURTURE

It is hard to avoid the conclusion that women are kinder and less aggressive than men. They are more likely to bear the brunt of caring for children and invalids, for example, and they are far less likely to be involved in violent crimes or warfare.

Despite this, there is no evidence that women are more altruistic in their dealings with strangers. In some ways, they are less so, although this can usually be construed as an unwillingness to take risks.

In general, there is little difference in altruism between men and women, as reflected in blood donorship and gifts to charity, for example. This may be because women are attracted to men who are generous to them. Hence the tradition of men footing the bill when couples go out on a date. Anthropologists believe that men, in some societies, may be willing to take care of unrelated children as a way of insinuating themselves in the affections of the mother so as to father her future offspring.[34]

Despite the general lack of sex difference in altruism, men and women behave differently in specific situations. For example, women are more likely to minister to the bodily needs and comforts of other individuals around them. The greater empathy of women for the discomfort of others is possibly connected to a greater personal sensitivity to unpleasant physical conditions. By the same token, the nurturance expressed by women in the care of children and sick relatives may be reflected in the much better care they take of their own health compared to men.

We do not really know why women are generally more nurturant in their dealings with family members and close friends, but there are many intriguing possibilities. One is simply the division of labor seen in all human societies in which women do most of the childcare. Feminists have fulminated about the unfairness of domestic inequality, but even feminist mothers still do most of the childcare, even if they are in full-time employment, suggesting that they are more sensitive to the needs of children or otherwise more motivated to care for them. Needless to say, there is great individual variation, and a minority of fathers are now the primary caregivers for their children.

How women grow up to be more nurturant toward children is currently a matter of some conjecture. There is no doubt that girls are more attracted than boys to playing with dolls and more interested in real babies. This would not be very compelling evidence for biological determination were it not for the fact that girls whose brains were masculinized by exposure to sex hormones before birth—whether due to a genetic disorder of the mother or exposure to a faulty birth-control drug—have little interest in dolls or babies and manifest a masculine toy preference for vehicles.[35]

Another interesting sex difference is in how children are treated by parents and other adults. From an early age, people play with males more vigorously and handle females more gently. It is difficult to know whether this behavioral difference is based entirely on what people believe is appropriate treatment for males and females, or whether male infants enjoy rough play more than females do. Parents not only handle their daughters more gently, but they also touch them more often. Are females more appreciative of physical closeness from an early age? Whether they are or not, presumably being the recipient of more physical affection predisposes females to provide such affection themselves as adults. (Females are not generally the subject of parental favoritism, of course, and in many subsistence societies, males are clearly preferred, often being allowed to eat before females, for example.)

Women around the world generally live several years longer than men, suggesting that they do take better care of themselves. In all developed countries, women live longer than men. In the United States, for example, the average man lives to seventy-three years, compared to seventy-nine years for the average woman. This difference may be partly due to biology: men have higher metabolic rates. When winter comes, women are more likely to complain of the cold before men do because their bodies produce less heat. One reason for this sex difference in metabolism is body composition. Men have more energy-burning muscle, particularly in the upper body, and women have a higher proportion of body fat, which would have been necessary in the evolutionary past to withstand the very great energy drain of pregnancy and lactation. Higher metabolism could produce more rapid aging in cells, thereby predisposing males of the species to earlier death. These ideas are still quite speculative, however. The fact is that there is currently no clearly established biological explanation for why men in most countries live substantially shorter lives than women. (Animal experiments *do* suggest that testosterone is important because castrated male rats live just as long as females.) There are, however, a plethora of well-established behavioral explanations. Men, in general, take more risks and are less willing to take good care of themselves by making health-conscious choices of diet, exercise, and lifestyle.

A man's most vulnerable organ is his brain! Masculine brains were designed by an evolutionary history in which men competed among each other over sexual access to women. This was not necessarily a trial of strength or physical courage, so much as a contest over social success, which is attractive to women. Modern men are willing to take grave risks to impress other men, thereby boosting their social success and desirability in the eyes of women. Males drive far more dangerously when they have other men riding with them than when they drive with women, for example. Unfortunately, risky driving is most prevalent in young men who are at a critical age for establishment of social status and dating desirability. Evolutionary psychologists believe that a great deal of the ostensibly motiveless violence of men in their teens and twenties is really designed to impress other men and so boost social status among peers and sexual desirability to women.[36]

Evolution has thus designed masculine brains to accept greater risks and discomfort than most women are willing to voluntarily undertake. Hence the overrepresentation of men in dangerous and dirty occupations. Risk-taking and failure to protect health are obviously cut from the same cloth, each being manifestations of an evolutionary history of male-male competition. The most direct evidence for this view is the finding that men who believe in traditional sex roles are less likely to go for regular medical checkups or to practice a health-promoting lifestyle.

There is no scarcity of data showing that men's health behavior is generally abysmal compared to that of women, although there has been a recent trend of increased drug abuse among young women, coinciding with a general convergence of gender roles in modern societies. Will H. Courtenay, of Men's Health Consulting in Berkeley, California, has assembled much of the evidence on sex differences in health behavior in a prodigious work of scholarship containing over six hundred sources. Here is a sampling of his conclusions:

- For all fifteen of the leading causes of death, men have higher death rates.
- Over one-half of young men (53 percent, aged 18 to 29 years) do not even have a regular doctor, compared to just one-third of women of the same age. Men make up 70 percent of individuals who have *not* visited their doctor any time in the past five years.
- Three-quarters of deaths from heart attacks of people under sixty-five are male. Many heart attacks are partly due to inherited vulnerability, but the risk of early death is exacerbated by unhealthy lifestyle, of which more below.
- Work life is a major risk for men, and 94 percent of work-related deaths are males. Why do so many men die on the

job? The simple answer is that they are much more likely to be killed because they accept more dangerous occupations, like fishing, mining, and farming. Men also tend to receive less sleep, which makes them more vulnerable to accidents and work-related injuries. Even being unemployed is more dangerous for men because of their vulnerability to alcoholism, depression, and suicide when deprived of a respectable way of making a living.

- One of the most telling statistics is that men are twice as likely as women to die from melanoma, a type of skin cancer. This is a fascinating statistic because more women sunbathe regularly (27 percent compared to 23 percent). Yet, men spend more time outdoors for occupational and recreational reasons. Despite their greater exposure to the sun, they are three times less likely to protect their skin using sun block. Either many men see this as a sissy precaution, or their inaction reflects a general lack of concern about physical safety and comfort—a feature of the male of the species due to inherited brain adaptations and differential treatment in childhood.
- Men pay less attention to eating a healthy diet than women do. They consume more fat and cholesterol and less fruit, vegetables, and fiber. They are less interested in having their cholesterol or blood pressure checked. Men are less likely to consume vitamin supplements that can improve health.
- American men are slightly more likely to be physically inactive in middle age than women are. Middle-aged women are 50 percent more likely than men to walk daily as a form of exercise. Men are drawn to sports like weight-lifting and football that have high rates of injury.
- Men are heavier smokers than women so that twice as many males die from smoking-related causes as females. This is considered the largest category of preventable deaths and is implicated in one-fourth of deaths for people between the ages of thirty-five and sixty-four.
- Men are more likely to be heavy drinkers. Three times as many men as women consume over five alcoholic drinks in a day,

which is the definition of binge drinking. They are over three times as likely to die from alcohol-related causes.

- Given that alcohol is a factor in about half of all homicides, it is no surprise that men are more likely to be involved in homicides than women. Homicide rates are particularly high among young men, and the bulk of these are described as "trivial altercation" murders in which a dispute begins over some unimportant matter, such as who is next to play on a pool table, and escalates to fatal aggression. Evolutionists interpret such disputes as really a conflict over social status, which has important implications for a man's sexual attractiveness to women. The point is well captured in Henry Kissinger's remark that "power is the ultimate aphrodisiac." Men are four times more likely to die in homicides. (They have a much greater risk of dying from suicide.)
- Alcohol is also a factor in drowning accidents, and men are five times as likely to die from drowning. For young men (fifteen to twenty-four years) the risk of drowning is ten times the female risk, reflecting a combination of risk-taking and alcohol use. Young men are generally more accident-prone and more likely to sustain injuries necessitating visits to the emergency room of a hospital.
- Risk-taking is also expressed on the roads, with 80 percent of injuries in road accidents being caused by male drivers. Two-thirds of road deaths are also men. Men are much more likely to drive while drunk and account for nineteen of every twenty drunk-driving charges. They are less likely than women to wear seat belts.
- Having handguns around is a major risk factor for serious accidental injury, and American men are much more likely than women to own guns (50 percent versus 22 percent). Males are more likely to keep their weapon loaded and are more likely to carry a gun. Men are seven times more likely than women to be injured by a gunshot and nine times more likely to die from a shooting.
- Men's greater mortality due to gunshots is partly a function of their involvement in violent crimes, with males constituting

95 percent of convictions for violent crimes, compared to 85 percent of crimes overall. Admittedly, these statistics may reflect greater leniency toward women in the criminal justice system. When a husband and wife cooperate in a criminal enterprise, for example, prosecutors often jump to the conclusion that the husband is the "real" criminal and devote their efforts to convicting him. Being shot at by police is a major occupational hazard of professional criminals. Police are four times as likely to kill men as women, reflecting, no doubt, a greater willingness among men to resist arrest.

- Criminals are also dangerous to themselves, at least in prison, where the suicide rate is sixteen times that of the general population. Interestingly, female prisoners do not have a high suicide rate, which might reflect a greater capacity for forming emotionally supportive relationships with fellow prisoners.
- In general, men are not as good as women at forming emotional support networks, and this has been identified as a factor in their vulnerability to leading causes of death. Men who are married, or have another close confidant, are three times more likely to survive heart disease compared to those who have no intimate friends. Married men enjoy much better health, probably because of healthier habits, including a better diet, better sleep habits, diminished use of alcohol and other recreational drugs, as well as whatever social support goes along with being part of a family unit.[37]

Medical researchers have thus assembled a staggering volume of evidence on sex differences in health behavior. Women take much better care of themselves, and their efforts pay off in the sense that they live longer, healthier lives. Although it is easy to jump to the conclusion that sex differences in mortality are a simple product of biology, the evidence suggests that behavioral differences are more important. Adding all of the behavioral risks together can account for most of the greater longevity of women. In other words, if men took as good care of themselves as women did, they would live almost as long. Why are men so reckless?

Evolutionary psychologists, as noted, interpret masculine risk-taking as a consequence of psychological adaptations for competing with other men, particularly over social status. Hence the bravado of young men in a variety of settings where they can impress peers, from dangerous driving to stunts pulled by skateboarders to valor displayed on the field of battle. Insensitivity to risk is particularly marked in young men, due, at least in part, to the effects of high levels of circulating testosterone on the brain. The strongest evidence for this is the fact that men are most prone to taking risks at an age when their testosterone level is highest. Violent crime, for example, is essentially an activity of young men between the ages of fifteen and thirty-five and is seen only at comparatively low levels in women and men of other ages. Testosterone acts on the brain, of course. Therefore we assert that a man's brain is his most vulnerable organ.

Men's peculiar lack of caution in terms of their health is counterpointed by women's greater interest in good diet, exercise for health reasons, and preventative visits to the doctor. If men's willingness to take unnecessary risks can be explained in terms of a history of mating competition, then the comparative risk aversion of most women might also have been affected by evolutionary pressures.

Women's risk aversion was likely favored by natural selection because women must look out for themselves if they are going to be around to protect and nurture their children. Among some subsistence peoples like the Ache of Paraguay, losing one's mother early in life is a virtual death sentence. Motherless children rarely survive, being highly vulnerable to malnourishment, disease, predators, and violence.[38] Other societies are more humane, and motherless children can be adopted by relatives. Men also play a role in the survival of children, of course, and it is interesting that as men approach middle age and as their testosterone levels decline, they become more risk averse. This more conservative trend is reflected in declining likelihood of being involved in road accidents and committing violent crimes, for example. For both sexes, it is possible that merely having the experience of protecting children from a variety of possible harms makes a person more risk averse because they must constantly think about possible sources of injury to children.

Risk-taking, as noted, provides social advantages for young men by raising their status among peers and thus increasing their sexual attractiveness to women. Unfortunately, risk-taking makes for poor health habits. Without invoking evolutionary strategies for maximizing reproductive success, medical researchers have no way of explaining sex differences in mortality. The best they can do is claim that the difference is "biological." In reality, there is not a simple biological explanation for higher male mortality rates. The difference *is* largely explainable in terms of behavior.

Men's irrational persistence in a great variety of risky behaviors may be partly due to differences in circulating hormones that affect the brain, but it is also affected by how they are brought up. Males are raised to be less nurturant toward others, and also toward their own bodies. Similarly, children raised in socially stressful environments, like crime-ridden inner cities, are socialized to be tougher, more willing to take personal risks, like committing crimes, than children raised in more comfortable environments. Even in such adverse environments, males are far more vulnerable than females. Their academic performance is worse. They are more prone to delinquency, more likely to join criminal gangs and succumb to drug addiction, accidents, homicide, and diseases associated with a reckless lifestyle. The riskiness due to a harsh social environment adds to the riskiness of being male.[39]

Nurturance early in life thus promotes careful treatment of one's own person as well as altruism toward other people. Merely behaving in nurturant ways can alter the biology of the nurturer too. This is one part of the jigsaw that needs to be assembled to reveal the complete picture of human altruism and its evolution.

NURTURANCE AND BIOLOGY: A TWO-WAY STREET

One of the most important discoveries of behavioral medicine is that merely being involved in a nurturant relationship improves survival following operations.[40] For example, pet owners are *four times* less likely to die in the year after cardiac surgery than patients without pets. This is a fascinating sidelight on the evolution of altruism that substitutes a win-win scenario for the Darwinian struggle among selfish individuals. In this case, by behaving in nurturant ways toward pets, cardiac patients derive immediate benefits. Being nurturant reduces stressful physiological reaction in humans as well as their animal companions, thereby improving the immune function and health of both.

Many of the evolved motivational mechanisms for parental care in birds and mammals are based on the simple principle of escaping unpleasant stimulation. When birds hatch eggs, they develop an inflamed patch of bare skin on their breast, known as a brood patch. The smoothness, and relative coolness, of the eggs is soothing. As the eggs heat up, they lose their capacity to soothe and are turned so that the cooler lower surface once again soothes the mother's breast. In this way eggs get periodically turned, thus avoiding the joint defects that result when eggs are left unturned in a mechanical incubator.

A conceptually similar motivational mechanism can be found in the distress calls, or cries, of newborn mammals that elicit nurturant responses in mothers. There are few sounds more aversive to human ears than the wailing of a newborn infant, for example. It evokes distress in others who are thereby motivated to relieve their own empathic distress by comforting the infant, at least if all goes according to nature's plan.

Such nurturant behavior is by no means confined to mothers, even in species for whom the female is normally responsible for all of the infant care. Among highly social species of mammals, it appears that the physiological responses underlying parental care are partially transferrable to other social relationships, or at least that is a reasonable working hypothesis to account for the effects of social support on health.

For many species of mammal, parental care is provided by mothers alone. (Fish fathers are the more nurturant sex.) This is true of rats, considered by biologists to be "conservative" mammals, that is, retaining many of the basic characteristics of the earliest mammals. The rat's scaly tail reminds us of reptiles that were entirely covered in scales, for example. Care of the young is performed entirely by females. Maternal caretaking activities include retrieving pups that have fallen out of the nest and licking them, a behavior that is essential for normal development, and even survival, of the young.

Maternal care in rats might seem like a genetically determined set of responses to helpless offspring. The truth is a great deal more complex. Rats must normally go through pregnancy and birth before their maternal tendencies are sufficiently primed to care for the pups. Virgin females will thus ignore pups that are placed in their cage. After several days of exposure to the infants, however, the females begin to take an interest in them and exhibit species typical care-taking responses such as retrieving and licking the pups. This is a remarkable example of the flexibility of animal behavior, but it is simply a question of inducing a female to perform her natural maternal behavior at an unexpected time. The really remarkable part of this story is that when adult male rats are tested in the same situation, they also retrieve and lick the pups following several days of familiarization with them.

Why would male rats have so much behavioral similarity with females if they never actually care for pups in the natural environment? Rats are highly social animals, frequently living in colonies of several hundred individuals. If social behavior of adults employs many of the same brain mechanisms as parental care, then the experimental unmasking of infant care by male rats may reflect a similarity between the brain mechanisms controlling infant care and those involved in adult social behavior.

Surprising though it may seem, there are many parallels between parental care in rats and in humans. Enduring the pain and stress of childbirth plays a role in the depth of attachment between human mothers and their newborn infants just as it does for rats. Thus, the level of stress hormones in the mother's blood before birth is predictive of the warmth of her relationship with the infant.[41] It is also clear that most mothers are far more devoted to their own infant than to others.

Maternal responsiveness to the crying of the baby has a distinct physiological component. There is an increase in heart rate and an increase in cortisol levels. The distress of one's infant is unpleasant.

The mother is physiologically aroused and ready to take action to relieve its distress. Maternal care thus fits in with the general features of other kinds of human altruism. Empathic arousal is unpleasant and altruistic acts are undertaken as a way of relieving the distress.

Most women, even those who have never given birth, are more physiologically reactive to a crying infant than most men. Even so, men who regularly take care of a baby produce exactly the same physiological responses as women. In a fascinating analogy with the rat experiments, actual experience in nurturant roles brings out a latent altruistic responsiveness to helpless members of the species. The only difference is that men play an important, if subsidiary, role in childcare in most societies, whereas male rats do not care for pups.

To some scientists, it may seem like a leap to equate the behavioral responses of rats to their young with the physiological responses of humans to babies, but there is a level of generality at which these can be seen as equivalent. All highly social species are, by definition, capable of a high level of cooperation, based on reciprocal altruism, or kin-selected altruism (or both). Such cooperation is based on physiological responsiveness to other members of the species. This is certainly true of human beings. Medical researchers have begun to amass a veritable mountain of evidence showing that lack of social support is one of the most serious risk factors for major killers like heart disease and cancer. Altruism is present in humans in the form of physiological responsiveness to other members of the species. We are physiologically calmed by the company of people we like or love, for example, which improves our health.[42]

SOCIAL SUPPORT AND HEALTH

Given the dramatic effects of touch on the growth of premature infants, it would hardly be surprising that social support should affect the health also of older children and adults. Much evidence supports this view. We know that social deprivation in the first year of life has devastating consequences for brain development as measured by infant intelligence tests and various measures of sensory function. Less well known is the fact that socially deprived children housed in institutions have a very high mortality rate.[43]

Even though their health and nutrition may be better than that of most other infants, institutionalized children, according to early researchers like Rene Spitz, had an unexpectedly high mortality rate of over one-third (thirty-four of ninety-one children) in the first year of life. For the first six months, babies raised in orphanages seemed quite normal. During the second half of their first year, however, they were uninterested in social contact and often appeared withdrawn or depressed. In addition to their many symptoms of retarded neurological development, institutional children were often stunted in their physical growth, just like the premature babies kept in incubators without tactile stimulation.[44]

One would imagine that the depression, and subsequent death, of infants deprived of extensive social contact is somehow connected to their very great physical and social dependence on adults as well. Yet, there are many cases of medically unexplained deaths among adults also. The contexts are extremely diverse. One of the strangest is the case of voodoo death, in which some member of a tribal society dies without any apparent bodily injury after being condemned by a tribal court. Reputable scientists, like physiologist Walter Cannon, have collected reliable accounts of such phenomena.[45] Of course this has nothing to do with voodoo as such and much to do with social condemnation.

Much nearer to the bone is the case of American prisoners of war held in China during the Korean conflict, where they were subjected to psychological torture. Many died while in good physical condition. Their deaths were connected to depression and were referred to by contemporaries as "give-up-itis." More recently, medical researchers have collected many accounts of people dying soon after the loss of a valued social companion, such as a husband or wife. It is difficult to draw any firm scientific conclusions based on such anecdotal accounts, but they do at least suggest that social support can be critical for health and even survival.[46]

Much of the research connecting health and social companionship has focused on cardiovascular disease, and cancer to a lesser extent, because these are the leading causes of death in developed countries where infectious diseases have been subdued through scientific medicine. Together they account for approximately two-thirds of deaths in the United States. Data implicating diminished social support in heart disease and other major causes of death include the following:[47]

- In a large Finnish study of bereavement, death rates for people who lost a spouse were approximately doubled in the first six months. Men were much more adversely affected than women. In Tacumseh, Michigan, and Gothenburg, Sweden, men were about four times as likely to die following loss of social support. Tacumseh women were twice as likely to die.
- Babies suddenly deprived of their mothers due to her death often refuse to feed and many die even if they are force fed.
- Men who stayed in Ireland, a country enjoying high social integration, had about four times fewer signs of heart disease than brothers who emigrated to the United States, despite a diet that was exceptionally high in saturated fats.
- Japanese men living in the United States who maintained ties to extended family and the local Japanese community in Honolulu, or San Francisco, had about four times less heart disease than those who had become fully Americanized and lost their Japanese roots, although dietary change could account for some of this relationship.
- Women with breast cancer who participate in group therapy live twice as long as those who do not.
- Among Harvard students in the years 1952–1954, who were studied thirty-five years later, 91 percent of those who had lacked a close relationship to their mother had serious medical crises by middle age compared to 45 percent of those who had been close to their mothers.
- In Alameda County, San Francisco, people who lacked either close social or community ties were twice as likely to die between 1965 and 1974. Women aged 60–69 were three times more likely to die in this period if they lacked social support.

- According to a large-scale study by the Centers for Disease Control, published in 1998, never-married people have a mortality rate 2.2 times that of currently married people. Widowed and divorced people have 1.8 times the mortality rate of married people.
- Unmarried white men in the United States are seven times as likely to die of cirrhosis of the liver as married men, eight times as likely to be homicide victims, and four times as likely to commit suicide.
- Married white male smokers aged 40–69 have the same mortality rate as divorced *nonsmokers* even though smoking usually increases the mortality rate by 100 percent.
- Cardiac patients without social support in a Yale study were almost three times as likely to die in the first six months as those with close friends and spouses. Similar results were produced in England and Sweden.

Literally hundreds of studies around the world have shown that frequency and quality of social interaction affects mortality due to cardiovascular disease and all other major killers. Needless to say, social support determines quality of life as well as longevity because the medical crises and chronic conditions that are more typical of people who lack social support are distressing and interfere with everyday activities.

Why should social interaction matter so much for cardiovascular health in particular, and other major diseases in general? At present, the complete story remains untold, but there are many intriguing clues concerning the hormonal basis of bonding and its effects on immune function.

Social animals get more relaxed in the company of individuals they know well and with whom they have a friendly relationship. Relaxation in a social context, and social bonding more generally, are mediated by two key hormones produced by the hypothalamus, namely oxytocin and vasopressin. Oxytocin is such a pervasive influence on social interaction involving physical contact that, as mentioned, it merits its nickname "the cuddling hormone."

In the human case, oxytocin is released by mothers when they nurse their infants, by children and parents when they hug each other, and by lovers in the throes of sexual passion. Oxytocin has many mood-enhancing effects. In addition to promoting calmness, relaxation, diminished sensitivity to pain, and a feeling of social connection, or oneness, the cuddling hormone affects immune function, helping to explain why close social relationships help us to fight heart disease, cancer, and other leading causes of death.[48]

Oxytocin is highly concentrated in the thymus gland, the major organ of immune function. Experiments on rats confirmed that the hormone has a direct anti-inflammatory action, which means that it can help to fight off the chronic inflammation that is a key player in the development of heart disease.

Like many other illnesses, including arthritis, cancer, and even Alzheimer's disease, heart disease is aggravated by stress and improved by social support. There is a very simple explanation for the role of social support because close social relationships boost oxytocin production, and the hormone acts as an antistress agent in the sense that it reduces the level of circulating stress hormones. This has been established in numerous animal experiments.[49] If oxytocin has an anti-inflammatory action, and if many chronic degenerative diseases are partly due to chronic inflammation, on which a consensus is just now emerging among medical researchers, then it is easy to grasp why social support protects against chronic illnesses like heart disease and contributes to good health and longevity.

Friendship, or affiliation, may thus be considered a primary need of humans and other animals, so that being involved in stable supportive social relationships has beneficial consequences for physiology and health. Research on the health characteristics of various communities backs up this claim.

COMMUNITIES AND HEALTH

Many authors have pointed out that loneliness is like an epidemic affecting modern societies in which frequent moves in search of better economic conditions break up long-established patterns of social relationships and create feelings of alienation.[50] Indeed, the research on heart disease among US immigrants from Ireland and Japan is a good example of this phenomenon.[51] Internal studies of community life within the United States also reveal some rather clear parallels between communal characteristics and health.

Particularly interesting is the contrast between Nevada, the gambling state, and Utah, home to the Mormon religion. These states are neighbors and they are broadly similar in their population structure. Yet, Utah residents have unusually good health and low mortality, whereas Nevada's mortality rate is above the national average. From 1994 to 1996, Nevada's death rate, of 553 per 100,000, was 26 percent higher than Utah's (410).[52]

What could account for this startling difference between neighboring states? Economics might matter if poorer people receive less adequate medical care, eat poorer diets, and exercise less, but Utah, the poorer state, has *lower* mortality. The Mormon religion in Utah may be an important factor because religion has been identified as an important source of social support among congregation members. In general, religious people, as determined by self-reported weekly church attendance, describe themselves as somewhat happier, and healthier, than those who do attend church less frequently, and have substantially lower mortality from leading causes of death, including heart disease, emphysema, cirrhosis of the liver, cancer, and suicide.[53] Given that twice as many Americans claim to attend church weekly as actually show up in church based on attendance records,[54] it may be that people in good health are just more willing to exaggerate their church attendance.

Mormons neither drink nor smoke, which also improves their chances of better health. Marriages are also comparatively stable, and people remain in contact with their families. Nevada, by contrast, has a very high divorce rate. Another important fact is that the great majority of Nevada's residents have originated from elsewhere, while much of Utah's population was born in state. All of this adds up to a more stable pattern of community relationships and social support. Better health habits and strong social ties may each contribute to the longevity of Utah's residents.

What is going on in an entire state may be difficult to study in a precise scientific fashion. For this reason, some researchers have chosen to look at small communities and track large numbers of residents in those areas. One example is the Framingham, Massachusetts, study of factors contributing to heart disease.[55] The problem of such studies is that they may be unrepresentative of the broader community. Thus Framingham had a much lower divorce rate than the rest of the country and could not be used to draw valid conclusions about the impact of marital disruption on health. Even so, research that is confined to small communities can reveal a great deal about the impact of social change on health. This is true, for example, of a study of Roseto, a small town in Pennsylvania inhabited by Italian immigrants.

Many of the immigrants arrived in 1882, having originated in the same small town in southern Italy near the Adriatic Sea.[56] Roseto was essentially a slice of Italian life transplanted to the United States. Residents kept many of the defining features of Italian life, cohesive three-generation families, devoted Catholicism, regular churchgoing, membership in social organizations, joyous communal celebration of traditional festivals, and rejection of the philosophy of social competition over wealth, or keeping up with the Joneses.

Roseto and its sixteen hundred residents first came to the attention of medical researchers with the recognition that its male death rate from heart disease was less than half that of surrounding towns, where heart disease was slightly more prevalent than the United States as a whole. The heart disease profile of Roseto inhabitants in their seventies resembled that of people in neighboring towns who were ten years younger. They were vulnerable to coronary heart disease, but their vulnerability seemed to have been magically delayed by a decade.

Like residents of Ireland, the people of Roseto did not have what the American Heart Association would describe as a healthy diet. They consumed a great deal of animal fat that is high in cholesterol. They also did relatively little exercise and had high rates of smoking. The combination of high rates of smoking and low rates of heart disease is also seen in Japan, a country that is distinguished by social cohesion.

With increasing prosperity, the population of Roseto began to leave their cramped old houses in the half-square-mile area of their town and built large new houses in suburbs. What they gained in real estate, they lost in community. With the physical separation of friends, it was no longer convenient for people to drop into each other's kitchens unexpectedly to share a cup of coffee.

The pattern of three-generational families has also been abandoned, and many of Roseto's inhabitants have married outside their ethnic group so that there has been a creeping Americanization, which meant increased materialism and decreased importance of religion. If community cohesiveness, and the social support it entails, were the fundamental cause of Roseto's low rates of heart disease, then the progressive Americanization of this slice of southern Italy would be predicted to increase the incidence of heart disease. That is exactly what happened. There was a steady increase in deaths from heart disease from the 1960s, when this village, sitting in the foothills of the Blue Mountains, first came to the attention of medical researchers, to the 1990s, when the mortality rate from heart disease caught up to that of neighboring communities. The correlation between changing lifestyle and changing health made this community of great interest for researchers interested in the connection between social isolation and heart disease.

The medical research on consequences of loneliness makes a very clear case that humans are a social species whose health suffers markedly as a consequence of social deprivation at any age. Given the substantial effect of social support in maintaining health, it is hardly surprising that most people should place a high value on stable friendships, marriages, and community ties. Yet, that is only the selfish side of the equation. Relationships are intrinsically give-and-take. They are mutualistic, and the partition of costs and benefits, beloved by modern Darwinians, is not really possible. If someone tells me about something that troubles them, and I listen attentively, then their physiological arousal is reduced and their immune system rebounds with declining stress levels. The interesting fact about this interaction is that the listener may well experience just as much of a health benefit from the interaction as the person venting his angst because listening, even to someone else's problems, reduces heart rate and blood pressure.[57]

This means that in interpersonal relationships, altruism is often difficult to define. Violently abusive relationships are clearly bad for a person's health, but most other benefits are shared by each of the people involved, although not to the same degree. Men probably benefit more from marriage than women, which could partly explain why women, particularly young women, are more likely to end unsatisfactory marriages. Conversely, women may gain more from same-sex friendships than men do, which may explain why they usually have a larger network of stable friends than men do.

If many social relationships are mutualistic, or win-win, then the problem of separate quantification of costs and benefits to each party is problematic. This is particularly true if the benefits of a social relationship are a matter of internal physiology.

The complex payoffs from social relationships are reflected in the underlying motivational psychology. Most people crave some kind of intimacy or social support, and the fulfillment of these needs contributes to their health and well-being. At the same time, some people, particularly parents, are strongly motivated to take care of their infants and children. Doing so probably contributes to the health of the apparent altruist—the parent. Nevertheless, it is reasonable to propose that parental behavior is a bona fide example of an evolved need to behave in altruistic ways. A mother may feel better when she comforts her crying infant, but this hardly disqualifies her behavior from being altruistic, as many cynics would claim. The effect of the altruism is more of a contribution to the health and survival of the helpless infant than it is to the mother herself. In that sense, it passes muster so far as the narrow biological definition of altruism used here is concerned. The mother's feeling good is the Darwinian mechanism through which maternal care is promoted.

Altruism to close friends and relatives is never entirely pure in the sense that there is a potential payback. If we build affectionate relationships with our children, they may stick around to cheer our old age and take care of us when we are helpless in our turn. Similarly, the fund of kindness that we invest in close friends may be drawn from at some time in the future when we need a favor. Some of the greatest acts of self-sacrifice are on behalf of close relatives and friends, people who are likely to reciprocate if they have an opportunity.

In the evolutionary past, most acts of altruism would have been directed toward relatives among whom we lived. The modern environment affords many examples of altruistic behavior, such as blood donation, that is directed toward complete strangers, however. Since there is no prospect of getting paid back in the future, such actions can be seen as unambiguous evidence of an evolved human need to help other people in their hour of need.

CHAPTER 7

Kindness among Strangers

Altruism directed at complete strangers provides the clearest case for an evolved altruistic motive in humans. This is a paradox because our evolutionary ancestors, living in small hunter-gatherer groups of fewer than one hundred individuals, were rather unlikely to encounter strangers because their social groupings were comparatively stable, and because population was sparsely distributed. We must therefore assume that when someone gives blood, or behaves courteously to strangers on the road, he is expressing a helpful attitude that would have benefited known individuals when it was expressed in the evolutionary past.

Altruism directed at strangers is a specifically human enigma for which there are no true animal parallels, although the cooperation among unrelated ant queens, which is widely interpreted as an example of group selection, comes close. Even here, there are two important differences. One is that the ant queens do not remain strangers. The other is that when one queen altruistically goes out to search for food, there might be payback. The others, for instance, may take care of her brood.

The human propensity to treat strangers kindly has recently been substantiated from a very unlikely source—economics. Economists have long clung to the assumption that humans are essentially motivated by their individual self-interest. That is the implicit basis of the notion of economic competition that forms the core of most economic theories.

THE ULTIMATUM GAME

Cutting-edge economic researchers use a variety of experimental games to investigate human altruism. These include role-playing exercises that mimic relations between employers and employees, and public-goods games in which players can make contributions to the common good. Such games are conceptually similar because they provide players with an opportunity to behave purely in their self-interest, as predicted by mainstream economic theories, or to behave with some degree of altruism.

The simplest of these experiments is the ultimatum game played by two people under conditions of anonymity. Their task is to determine how a sum of money, say $10, will be divided between them. One player, called the "proposer," makes an offer, which is to give the other player some amount between $1 and $10. The amount of the offer cannot be changed during the game. After the offer is made, the second player, the "responder," decides whether to accept or reject it. If the responder accepts, the money is shared accordingly. If the responder rejects, neither player gets any money.[1]

From the perspective of self-interested economics, the behavior of the players is easy to predict. The proposer knows that a self-interested responder will accept any sum of money greater than zero, so he or she should propose the minimal amount, $1. Yet, this self-interested outcome is almost never realized, and the average offer is nowhere close to $1. In fact proposers usually offer half of the money. Moreover, offers less than $3 are often rejected.

In the ultimatum game, it is thus clear that the behavior of both players is being governed by notions of fairness, rather than the pure pursuit of self-interest. Self-interested proposers should always offer $1, but they tend to split the money evenly. Self-interested responders should accept any offer, however small, but they turn down stingy offers because their sense of fairness is offended.

The responders' behavior of turning down the money is fascinating because they are literally turning down free money so as to spite a niggardly proposer. In other words, they are paying a price to punish the other player. Similar costly punishment behavior is observed in other experimental games and presumably reflects a generalized tendency to retaliate against exploitative individuals whose self-interested conduct follows the selfishness predicted by economics. The propensity for costly acts of punishment is referred to, particularly by group selectionists, as *strong reciprocity*.[2]

Evolutionary theorists have spilt a great deal of ink over how a costly tendency to punish selfish individuals could have arisen via natural selection. Most assume that the costs of punishment accrue to the punishing individual, whereas the benefits are distributed throughout the community. If so, then punishers are at a disadvantage. Leading theorists assume that although strong reciprocators are at a distinct disadvantage when they are thin on the ground, groups of reciprocators can prosper when groups of cheats fail.[3] This is a group-selection argument, however, and most evolutionists are unwilling to accept this scenario until all the possibilities of individual selection have been exhausted.

Punishment clearly makes no sense for the individual in a once-off game conducted under conditions of anonymity where the other players are strangers. Such conditions are extremely artificial, of course. Our ancestors were quite unlikely to have encountered complete strangers in one-time interactions.

The evolution of punishing tactics via individual selection is not nearly so puzzling as group selectionists imagine. Punishment can be considered a form of damaging aggression that has costs for the aggressor. Yet, no one ever claimed that *aggression* itself needed a group-selectionist explanation. It does have costs so that it is undertaken only when the likely costs exceed the benefits. Most animals settle disputes through a dominance system that allows them to avoid actual fighting. Yet, if the dominance hierarchy is disturbed, for example by the death of a high-ranking individual, they *will* fight because the prerogatives of elevated rank are worth it, as measured in Darwinian fitness (i.e., survival and reproductive success).

Just as an individual's place in the pecking order is worth fighting over, at least occasionally, so their desire to be treated fairly is worth standing up for. Not doing so, after all, has exactly the same consequences as loss of dominance rank, that is, a universal loss of deference in the society. Using this logic, it seems that theorists have been too willing to accept that punishment has no benefit for punishers.

The pattern of punishment seen in economics experiments is far more suggestive of individuals protecting their individual interests than it is of a mechanism to preserve a community of cooperators in the face of defectors. In general, players are more likely to use punishment when they feel that the other player acts in ways that threaten their vital interests, irrational though this may appear in the context of a one-shot game played between complete strangers. As already suggested, our ancestors lived in small groups whose members were very well known to each other. One-shot interactions would likely have been rare in the evolutionary past, and everything about the behavior of players in the ultimatum game suggests that they are offended when expected cooperation fails to materialize, that is, by the collapse of a sense of trust, rather than being motivated to exploit the situation for individual gain. Recent research on the sense of outrage in monkeys strongly corroborates this view.

In one amusing example of the sense of outrage in monkeys, Sarah Brosnan and Frans de Waal[4] of Emory University in Atlanta trained capuchin monkeys to exchange a token for food. The animals worked in pairs, each one getting rewarded in turn. If both received a piece of cucumber for their token, all was well. If one was given a grape, a more preferred item, the other refused to accept his cucumber, frequently tossing it right out of the cage in a fit of pique. There is a clear analogy with the actions of miffed humans in the ultimatum game. Annoyed by a sense of unfairness, they are willing to cut off their noses to spite their faces, metaphorically speaking.

Objecting to unfairness is a basic adaptive mechanism through which social animals assert themselves and protect their reputation. This conclusion is supported by research that examined brain scans of humans as they played the ultimatum game. When players received stingy offers, ancient regions of the brain that process emotions of anger and disgust lit up. Moreover, these intense negative emotions often had more of an impact on decision making than the more rational aspects of brain function, as reflected in cortical activity.[5] This is exactly what one would expect if punishment behavior in the ultimatum game is intrinsically emotional, that is, due to an aggressive outburst, rather than the more calculated decision required to punish a defector for the good of the group, as apologists for the altruistic punishment perspective (i.e., group selectionists) claim.

The ultimatum game is played differently in different societies. Players behave in ways that are consistent with self-interest in their particular social setting. Anthropologists have tested the game in various subsistence societies, producing different outcomes depending on normative patterns of sharing in the society. The comparison between two hunter-gatherer societies, the Ache of Paraguay and the Hadza of East Africa, was particularly striking. The Ache distribute meat with scrupulous fairness: the successful hunter gives the same amount of meat from a game animal to all other families as he takes for his own. The Hadza also share but their distribution of meat is characteristically tense and contentious, and hunters often attempt to conceal their catch from other members of the group.

Clearly, self-interest in these societies calls for different levels of cooperation, and this is seen in outcomes of the ultimatum game. The Ache always made generous offers: four-fifths were above 40 percent and no offer was ever refused, indicating a lack of spite in Ache sharing. The Hadza behaved very differently, however. They made low offers and many of the stingy offers were rejected, indicating a high level of punishment for selfishness. Note that this pattern is exactly the opposite from what would be predicted by a group-selection interpretation, where groups of altruists would be expected to bear the costs of punishment because it favored their group interest. Instead, one finds higher levels of punishment in societies that are low on altruism, exactly as the individual-interest interpretation would predict, but contrary to the altruistic punishment scenario.

There is no real surprise here, of course: exactly the same phenomenon is found in family environments characterized by low levels of altruism, such as those in which economic resources are strained. Developmental psychologists find that children in impoverished homes are more likely to experience spiteful behavior.[6] Needless to say, such spite has nothing to do with enforcing a group moral code—it actually undermines morality and altruism—and everything to do with protecting the individual interests of the actor.

The mere fact that punishing tactics undermine cooperation in some situations does not mean that they will do so in all situations, of course. Thus, people observe speed limits if they are liable to be fined. Similarly, public-goods experiments find that if defectors are punished, they are more likely to cooperate in the future.[7] Responding to contingencies of punishment has nothing to do with altruism, although group selectionists frequently claim that it does.

The group-selection interpretation of punishment sees the penalizing of defectors as a costly activity carried out by the individual for the benefit of the group. If so, then we would expect to see more punishment meted out by individuals in situations where the prevailing level of altruism is high. Yet, there is little evidence that this occurs and abundant evidence to the contrary. Of the many examples that could be listed, here are some striking counterexamples:

- Corporal punishment of children is much more common in societies characterized by a high level of adult violence.[8]
- Children of single mothers are at greater risk of severe corporal punishment reflecting family instability and reduced paternal altruism toward offspring.[9]
- Research on language development also finds that children raised in poor homes receive a great deal more scolding for misconduct, and a great deal less emotional support, than their counterparts raised by affluent parents. High levels of verbal punishment thus go along with low levels of verbal altruism.[10]

This example clearly illustrates the other side of the coin as well. Contrary to the predictions of group selection, punitiveness declines in highly altruistic situations. That is, affluent parents give their children much more social support, or verbal altruism, and much less scolding, or verbal punishment. Once again, there are numerous examples that could be listed of highly altruistic social situations inhibiting punishment:

- As already discussed, societies, and individuals, that enjoy a great deal of physical closeness in social situations are less aggressive.[11]
- Empathetic parents are less likely to use corporal punishment.
- Ambitious children come from families in which efforts to solve problems are supported, and where parents are careful to avoid criticizing children for their failures. Conversely, where parents are highly critical, children develop a strong fear of failure and tend to avoid situations where their inadequacies might be revealed.[12]

The vast majority of parents use some level of physical punishment on young children, at least where this is legal, and a case can be made that it is occasionally appropriate to inhibit actions that threaten a child's safety, such as playing with electricity outlets or running across the road without looking. Yet, the enlightened use of punishment, whether verbal, or corporal, is rare: much punishment is determined more by the emotional state of the parent than it is by the behavior of the child. In other words, punishment of children is largely due to an emotional reaction on the part of parents. Punishment is thus similar to other forms of aggression that can be explained in terms of individual costs and benefits without invoking any benefit for the community.[13]

In fact, a rather clear case can be made that corporal punishment, and verbal punishment as well, are *harmful* to the community. Here are some reasons:

- Punishment in childhood increases all forms of adult aggression from domestic violence to assaults outside the home and homicides.[14]
- Punishment undermines children's confidence in their own abilities and ambition with substantial economic consequences.[15]
- Punishment may interfere with the capacity of children to

> form satisfactory trusting relationships in later life. As a result, they are more prone to emotional problems, particularly anxiety and depression, predisposing them for drug addictions and a variety of health problems.

This analysis focuses largely on socialization practices of parents because we know far less about the emotional causes and consequences of punishment in adult situations, as reflected in experimental games in economics. The emotional dynamics of punishment in parent-child interactions are reflected in adult-adult interactions as well. Thus, societies with a great deal of religious, or ethnic, hatred are most likely to experience lynchings, ethnic cleansings, pogroms, and other atrocities. The urge to punish political enemies is born of personal hatred, not group altruism.

These relationships are particularly obvious in the case of outcast groups who are punished by the larger society for being who they are. Such unfair treatment does not encourage altruism even within the in-group. The following is a description of life among the Burakumin, Japan's untouchables, who were shut out of civilized society on account of perceived contamination by trades such as butchery and tanning that are abhorrent to many Buddhists: "Buraku men are allowed, if not expected, to be impulsive and physically aggressive, and the aggression is often directed toward their wives and children. Girls are spared no more than boys by their fathers. Mothers are generally more physically aggressive than they are in the majority society, especially than in the middle class. Fights between women are within the experience of most of our informants."[16]

Economic experiments are *highly artificial*. Thus, one-shot ultimatum games provide an intriguing glimpse into the assumptions that people make about fair exchange in their society, but they tell us very little about how altruism plays out over the course of numerous interactions. The public-goods game, in which individuals can donate some of their goods to enrich other players, is played in several rounds. As in the ultimatum game, players usually begin with the assumption of beneficence in others and make generous initial contributions to the public good. Over rounds, they become disillusioned with the contributions of others and curtail their own generosity. Once again, they can be seen as changing their behavior to protect their individual interests, curtailing their generosity when it falls on stony ground and fails to get reciprocated. One real-world public-goods "game" that is of enormous practical importance is the donation of blood. This is an intriguing problem in altruism because the healthy individuals who donate blood are particularly unlikely to need a transfusion. What is more, people in need receive blood regardless of their donation history.

WHY DO PEOPLE GIVE BLOOD?

Blood donation is largely voluntary in most countries. Voluntary donors are preferred over paid donors because their blood is safer, given that paid donors are more likely to have engaged in risky activities like injecting illegal drugs. The price of better quality blood donors is an uncertain supply. Even in a world of sophisticated medical technology, in which whole blood can be "stretched" in various ways, the prospect of imminent scarcity continually haunts blood banks and emergency rooms of hospitals. Blood is in constant demand for victims of serious injury and for operations in which there is a large loss of blood. Approximately one person in two will need blood at some point in a lifetime.[17]

One would imagine that the likelihood of needing blood at some point in the future would influence people's decisions about donating blood. This is like paying into an insurance fund so that when disaster strikes there is something there to cover the emergency. The problem is that there is no contingency between donating blood and receiving blood when it is gravely needed: people in need receive blood regardless of their donation history. The blood donation system is analogous to a car insurance system in which the premium payment is voluntary and the insurer always pays out in the event of an accident. No such business would be viable, of course. Yet, the voluntary blood donation system creaks along, a striking testimony to the strength of human altruism.

Before getting carried away by the workability of the system, it should be acknowledged that most Americans and most people in other countries never donate blood. They are essentially freeloaders in the system. Before asking why some generous individuals give blood repeatedly, it is worth asking why so many people refuse to contribute to a system that could be of tremendous benefit to them.

Hospitals rely on voluntary blood donation for accident victims and for performing many lifesaving surgeries. Blood donation requires minor inconvenience and discomfort but has no lasting ill-effects. Yet, fewer than 5 percent of eligible American donors give blood annually. Why are we so stingy about giving this precious gift even though most people engage in other kinds of charitable activity? For example, 90 percent of Americans give to charitable organizations each year, approximately one-tenth of them donating more than 5 percent of their income. About half of Americans do some kind of volunteer work.[18]

Researchers who want to increase blood donations are particularly interested in responses to this question. When interviewed, nondonors provide many plausible, although not necessarily reliable, reasons for failing to donate. Many claim that they are afraid of the pain and discomfort of being pierced by the needle used to withdraw blood. After a person has actually donated, fear of the procedure declines. When asked why they have ceased to give blood, former donors often point to lack of a convenient clinic, lack of free time, and minor health problems.

Such excuses may have some validity or they may be entirely face-saving. As far as convenience is concerned, for example, many people are given an opportunity to donate blood at work in mobile Red Cross units, but the self-employed may have no such opportunity. Donors have also been turned away in large numbers for a variety of medical reasons, such as being anemic, having difficulty finding a vein or yielding a full unit of blood, having had hepatitis, being homosexual, having cancer or heart disease, being exposed to malaria, or having had an opportunity to eat beef in England in the era of mad cow disease. Many potential donors are offended by the obligatory questionnaire that probes their personal lives in intimate detail. Have you had homosexual sex or injected yourself with drugs? Such questions are sensitive enough to turn some donors away, particularly if they are donating in a work setting and fear that their information might be "leaked." Given that the blood will be tested anyway, the questionnaire may also seem pointless. (Despite testing, initial screening is one of the most effective ways of ensuring safety of the blood supply; safety systems with built-in redundancy are better.)

Disqualification of many potential donors has hurt the US blood supply. In the case of severe local shortages, surgeries are delayed or even banned. Thus, the Red Cross recently declared an emergency ban on elective surgery in Georgia and even drafted restrictions on essential surgeries.

Perhaps the biggest reason that people do not give blood is that they do not see their donation as essential for the viability of the system. After all, they never gave before and hospital freezers are still full of blood. That is why the Red Cross wages a publicity campaign designed to convince people of the impending grave shortage of blood as donations decline and the demand for blood increases.

Scientists interested in boosting blood donations have collected demographic data of a type normally collected by market researchers in efforts to identify the typical donor. These efforts have not been particularly promising. Thus, in the past, the typical donor was male but the typical donor today is not. African Americans are less likely to donate blood than European Americans, but this difference may be due to the combined effects of poverty and a partly justified fear of the medical profession as hostile to blacks.[19] The appalling Tuskegee study in which doctors left syphilis untreated for decades to observe the natural course of the disease was conducted on poor African Americans, for example. The fact that blood was racially segregated in the early days of organized blood drives, in the context of World War II, has hardly helped African Americans to feel good about their blood donations.

The motivation for giving blood is not peculiar to any demographic group. The main motive is simply the altruistic desire to help others. Donors rarely cite egoistic motives but say that they are motivated by a sense of duty. They acknowledge feeling good about what they have done and recognize that being able to feel good about themselves was one motive for giving, although not the most important one.[20]

Blood donation is thus largely a question of people doing what they believe is the right thing. Yet, there is a definite emotional component. In situations where blood is desperately needed, for example if the life of a friend or relative is threatened and blood cannot be obtained quickly enough, or if there has been a disaster with a great deal of life-threatening injuries, most people will donate blood if they have the opportunity and if they are asked directly. This phenomenon was graphically illustrated after the terrorist attack on the World Trade Center on September 11, 2001. On the following days, huge lines of would-be donors lined up outside Red Cross centers in Manhattan. There are two reasons for this expression of solidarity with victims of the explosions in the Trade Center towers. One is that potential donors understood that large numbers of people had been injured by flying debris and needed blood. The second is that there was an enormous outpouring of sympathy for the victims and their families.

There is thus a very long list of potential reasons for failing to give blood and only a few reasons for being a donor, the most important of which is pure altruism. In an emergency, virtually anyone can be induced to give blood, but under normal circumstances, only a tiny minority of people volunteer. A behavior that is normal, and expected, in an emergency can thus be considered generous, or exceptional, under ordinary circumstances.

The altruistic explanation for blood donation is virtually unassailable because of the fact that there is no credible material return from such anonymous giving. Admittedly blood donation could conceivably have health benefits, but this possibility is not widely appreciated.[21] Donors are typically in good health and thus have no particular expectation of needing blood themselves in the near future. The contingencies are very different from the reciprocal food exchange systems of our ancestors in which all hunters had to contribute their game to the communal larder, so to speak, or risk immediate starvation themselves. Blood donation is thus a very different system, practically speaking. Nevertheless, it can be argued that the emotional adaptations for altruism that permitted our ancestors to share food may be invoked in modern situations where the calculus of costs and benefits is different. Human blood donation is not a reciprocal system of life-saving behaviors analogous to blood-sharing among vampire bats because one does not have to donate in order to receive, and donors may actually be less likely to get blood than others. Both of the key criteria of reciprocity are thus violated, that is, taking without giving, and giving without taking.

Giving blood is one of the clearest examples of disinterested altruism. Admittedly repeat donors may continue to give because they feel good about what they have done, but this cannot account for their first donation. It is also true that a person could make some social capital out of their status as a committed donor. The problem with this hypothesis is that blood donors do not characteristically seem to brag about their altruism, or even mention it.

Analysis of most other charitable activities suggests that there are mixed motives and that altruism is often fairly far down the list. People helping out in voluntary organizations often cite the following selfish reasons for getting involved:[22]

- They want to improve their job skills and/or experience to improve employment prospects.
- They want to gain exposure to potential employers.
- They hope to meet new people who will become friends or dating partners.
- They are interested in perks, like favorable publicity for themselves or travel opportunities.

According to the received wisdom of researchers in the field of voluntary work, volunteers are motivated more by self-interest than by concern for the people they are supposedly helping. Of course, that is a "social science," or philosophical perspective. To an evolutionist, the motivation is far less important than the potential fitness consequences for the volunteer. The fact is that voluntary workers give work time that could otherwise be devoted to income generation, for example by working overtime at their regular job. When they do so, they are very mindful of the potential benefits to themselves. The question of whether volunteers ultimately attain higher (or lower) social standing and income than nonvolunteers, by virtue of their service, is an empirical question that evidently remains unanswered. One clue is the fact that poor people are less likely to volunteer because they find the added expenses of travel and meals to be a problem. Volunteers may well be altruistic in an evolutionary sense even if their self-reported motives seem self-serving or egoistic. The biological definition requires only that the average costs of volunteering exceed the benefits. This would appear to be true in most cases even if there are occasional instances of a volunteer coming out on top by landing a better job, for example.

It is also possible that blood donation, for which self-reported motives are most often altruistic, does not involve any fitness cost. If people feel good about their donations, this could increase their sense of community involvement, which is known to have a protective effect on cardiovascular health. It would be fascinating to examine the health of donors as a function of their frequency of donation. Do frequent donors have better health and longevity? Since an evolutionary approach to helping behavior is still in its infancy, it seems there are currently few answers to such fascinating and, in practicality, important questions. Research conducted at the Mayo Institute suggests that blood donors are at a reduced risk of heart disease and stroke, but it is not possible to say if this is due to giving blood or to other attributes of the sort of people who are donors, such as social support, diet, and exercise. If donating blood turned out to have health benefits, this would be a fascinating vindication of the most longstanding surgical practice in the history of medicine, now thoroughly discredited, namely blood letting, which had been practiced for some 2,500 years.[23]

Whatever about the precise economic and personal costs of volunteerism, there are some situations in which people have voluntarily sacrificed a great deal to protect the interests of strangers, or at least nonrelatives. From soldiers on suicide missions to firemen entering a dangerous burning building that may collapse at any moment to the daily risks assumed by fishermen, construction workers, and bodyguards, there are varying levels of heroism, but the bottom line is that the rest of us benefit in some way from the self-sacrifice of brave people, whether such altruism is motivated by financial gain or by the purest of heroic sentiments. One example of personal sacrifice that has received careful scholarly attention concerns those individuals who risked their own lives to shelter Jews in Nazi-occupied Europe. This attention is thanks in part to the efforts of Jewish organizations that wanted to remember and honor their heroic benefactors.

HOLOCAUST RESCUERS

Heroism is often simple in practice but complex in principle. Take the case of six soldiers taking cover in a shell crater when a hand grenade is lobbed into their temporary refuge. If no one takes action, all of the soldiers will be seriously injured and will likely die. If one is willing to sacrifice his life, he can cover the grenade with his own body, thereby blowing himself to bits, but blocking most of the shrapnel from reaching his companions.

This tragic circumstance is like a one-trial six-person prisoner's dilemma. Game theory would predict that all the soldiers should do nothing because that is their better choice in a very bad situation. Yet, there is evidence of soldiers being willing to sacrifice their own lives to save their companions. In fact, among 207 soldiers honored with medals of honor during the Vietnam War, fully sixty-three distinguished themselves by using their own bodies to shield comrades from exploding devices, mostly live grenades. Sociologist Joseph A. Blake of Virginia Polytechnic Institute studied these heroes and found that their suicidal altruism was motivated by loyalty to their fighting unit.[24] Soldiers were most likely to sacrifice themselves if they belonged to highly cohesive units (assuming, perhaps controversially, that elite branches of the military are more cohesive than nonelite branches) and were motivated more by loyalty to their comrades than to their military objectives as such.

Blake also found that sacrificing oneself on a grenade was more common among enlisted men than among officers, although officers received proportionately more medals of honor for other types of heroism in combat. A likely explanation is that the small group of enlisted men in a fighting unit builds up more cohesive relationships among each other than is possible between officers and enlisted men. The survival of the entire unit may also be dependent on the heroic cooperation of each of its members, so that no personal sacrifice is too great.

The important question to ask in such situations is, "Why does a particular individual assume the role of hero?" Most of us are willing to admit that some people are just more heroic than others. If a man gets faint at the sight of blood, for example, he may be unlikely to be the one who covers the grenade.

There is clearly a personal component, related to a person's fearfulness and his capacity to override instinctive reactions of self-preservation. There is also an important situational component. The soldier who happens to be crouched nearest to where the grenade lands is particularly likely to cover it for two reasons. First, he has the best chance of getting to the grenade before it explodes. Second, he is likely to be severely injured or killed, even if he fails to act.[25]

People are prone to overestimate the importance of character in such decisions and to underestimate the role of the situation. This error is so prevalent in social cognition that psychologists have come up with a fancy name for it: the fundamental attribution error.

The general question of why people help strangers was addressed by Samuel and Pearl Oliner of Humboldt State University in their study of individual Christians who rescued Jews in Nazi-occupied Europe. The Oliners based their conclusions, presented in a book, *The Altruistic Personality*,[26] on interviews with over seven hundred rescuers (i.e., those who helped Jews) and nonrescuers (who didn't help). Rescuers were identified and carefully authenticated by Yad Vashem, a Jewish organization dedicated to commemorating the Holocaust and honoring Christians rescuers. Over six thousand such individuals have been identified. Even though the Oliners focused on the personality factors of their altruists, their research also uncovered some interesting situational determinants of helping. Thus, people with cellars in their homes were more likely to be rescuers, evidently because cellars provided a convenient refuge where extra people could be hidden. For some reason, having an attic made no difference. Rescuers were also better off, so that they had more prospects for feeding extra mouths during the economically tough times of the war, and they had larger homes, that made it easier to accommodate Jews. There were many other situational determinants of who helped, such as actual or perceived risk of being found out by the Nazis, or being turned in by informants. Having a specific opportunity to help was important. Thus, Christians were more likely to help if Jews had directly asked for their assistance.

In making up their comparison group of 123 nonrescuers, the Oliners uncovered an interesting, if inconvenient, fact. Even though they had not shown up on Yad Vashem's list, 42 percent of the control group claimed that they had either helped Jews in some way during the war, or had worked in resistance movements, or both. Faced with this unexpected twist, they had to divide the nonrescuers into actives and bystanders. Although these self-reports of resistance activities could not be validated and might have contained self-serving biases, they nevertheless indicate that less heroic forms of self-sacrifice than those shown by the comparatively small number of rescuers may actually have been more widespread.

That would help to explain the common perception among rescuers themselves that they had done nothing out of the ordinary. They felt that they had done what any decent human being would have done if placed in the same situation. In other words, there was a widespread sentiment in the occupied territories that people should do whatever they could to resist the Nazis and their insane policies against the Jews. Most people evidently did not live up to this belief. Some were active in the resistance, and a handful risked their own lives and those of their spouses and children by sheltering Jews in their homes, often for several years.

Although the rescuers may not have felt that they were unusual people, the Oliners' study suggests that many had unusual parents. Active altruism, of the kind that is seen also in political activists, is unusual, and one of its strongest determinants is parental example. Remember that of the deeply committed civil rights activists studied by David Rosenhahn, most had activist parents. Similarly, rescuers of Jews had witnessed examples of heroic kindness by their own parents. Many recognized that their parents would have behaved in exactly the same way as they did. From this recognition, it is perhaps a small step for them to conclude that any decent person would behave in the same way.

Situational influences can be extremely important determinants of helping behavior. Social psychologists subsequently established this through many ingenious and controversial experiments (see next chapter). Given the importance of the situation, we might imagine that many of the rescuers knew the Jews they helped rather well. According to the Oliners, the shoe was on the other foot: they concluded that 90 percent of the rescuers helped at least one person who was a stranger. This is an important fact because it suggests that the decision to help was often made on ethical, rather than personal, grounds. There was, nevertheless, an important experiential component: rescuers were more likely to have had Jewish friends before the war, to have had Jewish coworkers, and to have had Jewish neighbors. Two-thirds of rescuers were also asked for help at least once.

As families shared the same living space, they often became quite emotionally attached to each other, so that personal considerations became more important as time went on. In some cases, as conditions became more dangerous, Jews wanted to leave, or even turn themselves in to the Nazis, to avoid visiting destruction on the homes of their benefactors. In most cases, rescuers refused to let the Jews go, even though this would have let them off the hook. It is not difficult to imagine why they should react in this way. Apart from the developing bonds of friendship, exposing their charges to the increased risk of being captured when outside their refuge would have undone all the sacrifice invested in them thus far.

This point was made forcefully by a Dutch rescuer whose guest, a young Jewish man, wanted to turn himself in, rather than endanger his host, after they heard that the house was to be searched:

> I was so angry with him. Can you understand that? I said, "You have been here for over two years. Now you want to give up? Are you crazy? Take off your coat and sit down. Listen to me and listen carefully. On a ship there is one captain. I am the captain, and you have to listen to me. I will find something. I always find something. I have not gone to all this trouble in order to fail."[27]

This Dutchman was clearly domineering as well as fearless, traits that were common among rescuers.

Most of the rescuers interviewed in the Oliners' study felt proud of their actions and believed that they had done some good in the world. The altruistic action may be its own reward. One respondent commented:

> I did everything from my heart—I didn't think about getting something for it. My father taught me to be this way. I feel the same way now. I cannot refuse if someone needs something. That's why I still help people—I'll do it until I don't have the strength to do it anymore. My nature is the result of being raised by my mother. She was my role model. She helped a lot of people.[28]

Most of the rescuers felt that their actions were quite natural and ordinary, that they lived up to the normal human desire to help others in desperate need. Whether this analysis is correct or not, the rescuers were statistically rare. They provide a unique opportunity to investigate whether heroic helping improves people's health, perhaps by making them feel more socially connected. The Oliners' data did not show that rescuers were healthier, however: if anything their health was somewhat worse.

Although heroic altruism toward the Jews was not common, the great majority of Europeans sympathized with their plight. For example, 86 percent of the nonrescuers in the Oliners' study felt empathy toward the Jews the first time they saw a Jewish man or woman wearing a yellow star, which was not statistically different from the 92 percent of rescuers who had this reaction.

The modern world is full of opportunities to help strangers that could not have existed in the evolutionary past, such as donating blood and sheltering victims of an oppressive government. Adoption of complete strangers through modern adoption boards is another intriguing example of a behavior that is puzzling to evolutionists because it diverts parental investment away from biological kin.

ADOPTION AS ALTRUISM TOWARD STRANGERS

According to evolutionary theory, adoption may seem like a very puzzling behavior because of the diversion of resources away from kin to raise complete strangers. Yet, it is a mistake to see evolutionary processes as always producing individuals who maximize their fitness. Thus, as already described, many societies have holy people who renounce sexuality and reproduction for most, or all, of their lives. From a purely reproductive perspective, natural selection in the past did not protect these unusual individuals from the kind of current social environment in which they accept celibate social roles.

Human beings, and other animals, are designed by natural selection to behave in ways that contribute to their reproductive success, just as engineers design planes to fly through the air. People usually reproduce if they have the opportunity and planes usually fly. There are occasional glitches, however, so that some individuals do not or cannot exploit reproductive opportunities. Planes usually fly but occasionally they fall out of the sky. However good the design, there are unusual conditions in which machines and organisms do not do exactly what they were designed to do.

Accounting for adoptions of complete strangers has two complementary pieces. The first answers the question "Why adopt?" and the second "Why adopt strangers?"

The desire for children of one's own is a recognized human need that is separable from other aspects of the physiology of reproduction, such as sexual desire. To some extent, this need is based on nurturant tendencies. Developmental psychologists find that from an early age girls are more interested in dolls as playthings and more motivated to interact with real babies than boys are, and these sex differences are evidently affected by prenatal exposure of the brain to sex hormones. In preindustrial societies, girls spend far more of their time looking after younger siblings than boys do. Despite such sex differences, some women are not particularly interested in babies until they produce one of their own, and the reactions of fathers are often similar.[29]

As to why adopt a child in the event that a couple is infertile, there are at least three plausible psychological reasons. One is that raising a child is helpful to a marriage, giving the partners a shared interest and shared investment in the future, as well as the joint commitment and responsibility of raising a child. Childless couples tend to drift apart and have an unusually high probability of divorce. According to UN data, couples with no children are twice as likely to divorce as those having two children, and ten times as likely to divorce as those having four or more children.[30] The second is that some people, particularly women, have strong nurturant tendencies toward children. The third is that babies were designed by natural selection to be successful at tapping in to this nurturant tendency of adults. Early behavioral biologists pointed out that all juvenile mammals, and even birds, share typical features that make them attractive to caregivers. Whether you look at a young pup or a human baby, you find a generally rounded shape, a head that is large in proportion to the body, comparatively large eyes, and a shortening of the lower portion of the face compared to the upper part. As mammals age, all of their bones lengthen, including facial bones, so that there is an increase in the impression of angularity as opposed to roundness. Austrian behavioral biologist Konrad Lorenz referred to the typical infantile appearance of the young as a *kinderschema*. In everyday terms, we call it cuteness. Cuteness is a major reason that human families adopt young pups in preference to older, better trained, animals. Also, they'll likely be around longer.

Biologists refer to such cross-species exploitation of parental care as brood parasitism, which makes it sound rather nasty. Interestingly, the exploitation is not always a case of dogs pushing people's buttons. There are several reasonably well-described cases of human infants being raised by nonhuman mothers (see fig. 7). Such feral children have evidently been raised by a variety of hosts, but wolves are often cited as the protectors of human infants.[31]

Given that human adults and children have inherited predispositions to fall under each other's spell for the purposes of preservation and care of the infant, the success of most adoptions is a foregone conclusion, particularly if the adoption occurs early in the child's life, ideally in the first few months.

People are predisposed to care for babies left in their charge because of psychological adaptations for care of their own children. The fascinating aspect of adopting strangers is that adoptive parents behave toward their adopted children almost exactly, and in many cases exactly, as they would behave toward their own offspring. If this is true, and there is some disagreement on the details, as noted above, they are violating a key prediction of kin-selection theory, namely that individuals should invest their parental efforts preferentially in offspring, or close relatives.

Why adopt strangers? The phenomenon of a pool of adoptable

Fig. 7. Cross-species adoption, as depicted in a statue of Romulus and Remus, the legendary founders of Rome. (Library of Congress)

babies is a modern one. In the past, if mothers could not hope to raise children successfully, they often killed them at birth. Infanticide is still something of a problem, particularly among young, poor, and ignorant women, but adoption societies rescue many infants that would otherwise have perished because of dim prospects for economic support. In the evolutionary past, when mothers died suddenly, their dependent children were likely taken care of by close relatives. All of this means that unattached, helpless, strange infants were not a feature of the social environment inhabited by our hunter-gatherer ancestors. Because there was little possibility of expending parental effort on unrelated infants, natural selection did not equip us with a mechanism for reliably detecting closely related infants and refusing to care for strangers. It could easily have done so. Thus, people have some limited capacity to recognize the bodily odor of close relatives. This ability is not very reliable, however, because babies occasionally get mixed up in hospital nurseries, a mistake that is often not discovered until the children grow up and turn out to be biologically incompatible with both of their parents.[32]

In recent years, popular attention was captivated by two highly publicized cases of baby switching. In one case, two little girls were switched at the Charlottesville, Virginia, hospital where they were born. This came to light when one of the mothers, Paula Johnson, went to court in an effort to compel her ex-boyfriend to resume child-support payments. Court-ordered DNA tests revealed that the little girl, Callie, who was the subject of the dispute, was not biologically related to either the woman or the man. Their real daughter had been living with Whitney Rogers and Kevin Chitum, who had raised her as their own and named her Rebecca.[33]

The second case came to light in 1989 when it emerged that ten-year-old Kimberly Mays was switched with another little girl in a Wauchula, Florida, hospital. The other girl subsequently died of a congenital heart defect. The biological parents, Ernest and Regina Twigg, sued to obtain custody of Kimberly Mays, but she remained loyal to the couple who had raised her and ultimately won a suit that denied parental rights to the Twiggs. Such confusions attract a lot of publicity but they are quite rare because hospitals use a careful tagging system. Matching identification bands are attached to the mother's wrist and to the wrist and ankle of the infant soon after birth, and these are checked carefully and often. The fact that parents fail to detect such switches suggests that human kin recognition is unreliable in the case of babies. When a lack of relatedness is unknown, parents may treat a complete stranger exactly as they would treat their own child.[34]

The capacity to adopt complete strangers can, however, be interpreted as a consequence of adaptations designed for the nurture of biological offspring. To an evolutionist, adopting unrelated children is intriguing because it violates the expectation of parental investment in close kin. Yet, the capacity to nurture completely unrelated children is a fascinating example of universal human altruism that is truly fortunate for those children who can benefit from it. The only major theoretical question left to ask about adoption is whether being raised by adoptive parents is as good as being raised by natural parents. Do adoptive parents invest as much in their adoptees as biological parents do in their offspring? If you compare adoptees with children raised by their natural parents, do they differ as adults in ways that reflect varying levels of parental love and nurture?

One study that addressed this question was based on the National Longitudinal Study of Youth, a large and representative survey. Large as the survey was, reaching over ten thousand people, it still contacted only 101 adopted children because adoption is comparatively rare, affecting only about 1 percent of children. On some measures of adult success and life satisfaction, adopted children, in their late twenties and early thirties, were indistinguishable from those raised by biological parents. They were not more likely to have used drugs recently. Educational attainment was the same, as were occupation level, job stability, probability of home ownership, and asset accumulation. Adopted children were not more likely to divorce than children raised by both parents. Yet, among this small sample, female adoptees were more likely to have cohabited before marriage and to be less happy in their marriages.[35]

There is an odd inconsistency in the fact that even though it is difficult to distinguish between adult adoptees and children raised by both biological parents, earlier in life, during adolescence, adoptees have a higher incidence of using alcohol and illegal drugs, more delinquency, and higher crime rates.[36] They are also more likely to suffer from depression. Some of these problems last into early adulthood but they are evidently cured by time. Interestingly, the problems of adopted children are virtually indistinguishable from those of children raised in stepfamilies or in blended families (in which both previously married parents have children).

The greater risk of suffering from depression among adopted children means that they are more likely to attempt suicide (7.6 percent versus 3.1 percent of all other children, according to data from the National Longitudinal Study of Adolescent Health).[37]

Reading between the lines, it appears that adolescence is a particularly difficult time for adoptees, and there is evidence that they feel less connected to their families. Could the emotional problems reflect a lack of emotional commitment from their adoptive parents and adoptive siblings, or might it simply reflect the alienating consequences of finding out that Mom and Dad are not one's "real" parents, that the real parents abandoned one soon after birth (as many children would be likely to interpret the situation)?

Do parents invest less in adoptive children than they would in biological children? This is a delicate question because it seems critical of people who sacrifice so much to help strangers. It is thus appropriate that the unkindest cut should come from economics, a science with a vested interest in believing that people are driven entirely by selfish motives and which actually indoctrinates undergraduates into becoming more selfish.[38] It turns out that adoptive families spend less money on food in the United States than do families that include both biological parents. The same reduction in food expenditures is seen in stepfamilies and families with foster children. It would be easy to write this study off as some kind of statistical fluke, but it has been replicated in South Africa. The South African study found reduced expenditure on milk, fruit, and vegetables, and increased expenses for tobacco and alcohol. Economist Anne Case of Princeton University, the first author of the research, sees the findings as evidence of "hungry" selfish genes.[39] Yet, there is another plausible interpretation. Nontraditional family types can be more stressful, prompting greater expenditure on self-medication for anxiety in the form of alcohol and tobacco. Since disposable income is fixed, the food budget suffers from increases in the anxiety-reduction budget.

Such uncritical selfish-gene arguments have been around for many decades in the field of evolutionary psychology. Canadian researchers have reported the rather grisly fact that children in stepfamilies are more likely to be homicide victims and are much more likely to be killed by stepparents than by biological parents.[40] Stepparents also treat their own children more favorably in other respects, as already discussed. The Canadian data are of interest to criminologists who deal with very unusual individuals but may have little relevance for how the great majority of adoptive parents treat their children.

Stepfamilies are not an ideal test of parental discrimination against unrelated children because they have so many sources of emotional conflict compared to biological-parent-only households. One, or both, partners has already had a failed marriage, suggesting inability to deal peacefully with conflicts in intimate relationships, for example. There is also the difficulty of forming a bond with an unrelated child after the age of infancy. A much less loaded comparison would be to contrast death rates from homicide of adoptees with children in families containing both biological parents. Adoptees do not appear to be at greater risk of homicide than biological children. This is no surprise if people are biologically predisposed to bond with children they take care of from an early age. If so, the parent-child relationship should be virtually indistinguishable whether the child is adopted or not. Just as biological parents would be loath to injure the children they have nurtured, so adoptive parents would do everything in their power to further the interests of their charges.

Altruism toward strangers can thus approximate the huge investment of parents in their biological children. This is superficially contradictory of evolutionary predictions until one realizes that our ancestors almost never confronted the problem of live, abandoned and unrelated babies throughout the millions of years of hominid evolution.

Up to this point, all of the examples of substantial altruism toward strangers have been voluntary. We also contribute to strangers when we pay our taxes, although this is largely a matter of compulsion. Paying taxes is superficially very different from caring for adopted children, but both are instances of living up to complex obligations in respect to nonrelatives and may thus tap similar evolved altruistic motives.

PAYING TAXES HONESTLY

Paying taxes is usually not voluntary for individuals, although corporations can wiggle out of their legal responsibilities by various strategies of relocating to tax shelter countries or creatively shifting profits to countries with the lowest tax burdens. Despite its compulsory nature, personal taxation does have voluntary elements, at least in countries like the United States where individuals must prepare complex tax returns that can never be completely audited by the government, which uses the uncertainty of random audits to encourage honest tax preparation. In many European countries, by contrast, tax obligations are figured by government accountants and extracted directly from paychecks, which gives individuals little opportunity for tax evasion.

In the United States, taxpayers must prepare complex tax returns that frequently include ten or more forms, each of which is based on a voluminous maze of regulations and exceptions. Professional tax accountants are skilled at finding obscure deductions that reduce the tax bills of individuals and, particularly, of corporations. Some exemptions are available to everyone. As recently as just over a decade ago, individuals could claim charitable deductions of $3,000 without receipts. Almost incredibly, few Americans were willing to exploit this loophole until it was closed in the late 1980s. They also rejected more extreme types of tax evasion. Thus, some people pay no taxes at all, claiming that key portions of the tax code are vague and unenforceable. They file zero-income returns and answer

Internal Revenue Service (IRS) queries with carefully drafted letters. Tax experts believe that such activities are illegal, but the fact remains that some individuals have paid no taxes for as much as ten years despite being taken to court by the IRS.[41]

Given the many ambiguities of the tax law, from its intricate details, such as which depreciation schedule and convention are appropriate for writing off a particular type of business asset, to whether a person is legally obliged to file accurate tax returns, one would imagine that large numbers of Americans would be comfortable in exploiting the regulational fog for their own benefit. Yet, in 2001, over three-quarters of Americans (76 percent) responding to a survey said that they believe in being completely honest on their taxes. This number was a significant drop from the 87 percent of respondents who believed in complete tax integrity in 1999.

The decline in beliefs about tax honesty could reflect publicity given to tax-evading individuals and corporations. Paying taxes honestly is rather like a prisoner's dilemma game. Tax revenue is ideally used by governments to promote the common good, so that everyone is better off. If most people cooperate, meeting their tax obligations honestly, then a few tax-hating survivalists can be tolerated. As soon as ordinary people perceive that others, just like themselves, are cheating, then defections become more likely.

Paying taxes honestly when it would be easy to cut the tax bill using questionable deductions, or even to pay no taxes at all, can be thought of as a form of altruism directed toward strangers. The strangers are not just governmental administrations, of course, but the ultimate beneficiaries of government disbursements, including public employees, government contractors, Social Security, Medicare and Medicaid recipients, and others who benefit from government spending.

Given that most people believe tax payment is compulsory, describing it as altruism may appear to stretch the definition fairly thin, even in terms of a narrow biological interpretation. Nevertheless, disbursement of taxes is a transfer of resources from individuals and their family to strangers. If most people behaved in conformity with expectations of biological selfishness, they would be willing to fudge their taxes in order to benefit themselves and their families. Yet, the great majority are not willing to cheat on their taxes, just as the majority do not commit crimes when they have the opportunity to do so.

In part, tax honesty may be related to fear of being caught evading taxes and being legally compelled to pay back taxes, with interest, and severe penalties. Thus the decline in beliefs about tax honesty from 1999 to 2001 followed a sharp drop in IRS enforcement in the aftermath of a scathing 1998 investigation into abusive enforcement by IRS agents. By 2000, the IRS was examining only one in 232 returns, compared to one in 79 for 1988. Despite such declines in enforcement, the majority of people believe in filing honest tax returns, just as they believe in supporting worthy charities. They always have, and, hopefully, they always will.

Far more people express their altruism toward others by filing honest tax returns than engage in more intimate acts of self-sacrifice, like giving blood. This partly reflects the compulsory aspect of tax payment, perhaps, but there is also an element of social obligation that transcends legal compulsion. As good citizens, we have a social obligation to file honest tax returns, but giving blood is both personally and legally optional.

Honesty in taxes is part of a pattern of honesty in other spheres. Researchers discovered that company officers who cheat on corporate taxes are also more likely to cheat on their own taxes. One study even found that dishonest executives are more likely to cheat at golf. The majority of people want to see themselves as being honest because of our evolutionary history of honest dealing in alliances with known individuals. We tip waiters in distant cities because we believe that this is the correct thing to do. We obey laws and do our best to file honest tax returns.

Filing accurate tax returns is an extremely complex example of the human ability to follow shared rules of expected conduct. Such conformity is the foundation on which a great deal of altruism among strangers is built, from our ability to observe rules of the road, and rules of dress, to the acceptance of shared belief systems. Perhaps the great care that adoptive parents take of their children also reflects, in part, at least, conformity to the rules of what is expected of people in this role.

CHAPTER 8

Conformity as Altruism

Put people in uniforms and they look pretty much the same. What is more remarkable is that they also behave similarly and direct their actions toward common goals. The human capacity for fitting in with the demands of group living not only is responsible for the tightly coordinated actions of military units, but also facilitates the much more diffuse coordination of behavior that allows modern cities to function rather like giant anthills. One conspicuous example of fitting in is the rules of the road, which are essential for safe and efficient travel (see fig. 8).

People subordinate their actions, and even their ideas, to the requirements of social groups and authority figures, according to copious research by social psychologists who have paid less attention to explaining why humans act so slavishly conformist and obedient. Human groupishness, or conformity, is seen at its most extreme in complex societies, but the underlying psychology evolved so that we could function effectively in small kin groups. Even in hunter-gatherer societies, foraging groups are quite fluid, however, and they are combined in tribes that may number in the thousands.

Conformity to the wishes of others is altruistic because the immediate interests of the individual are sacrificed to the benefit of the group. Individuals who worked well in small groups were more likely to survive and reproduce, of course, because the group was more successful at obtaining vital resources, such as food, shelter, and security

Fig. 8. Lane discipline: safe driving requires elaborate conformity to rules of the road, written and customary (see text for explanation). (Library of Congress)

from enemies and predators. This explains why groupish tendencies evolved through natural selection acting on individuals.

CIVILITY AND EVOLUTION

Civility is a much-valued component of communities that is frequently lost in cities because large numbers of strangers are pressed together in anonymous situations. Simply greeting strangers can make them feel part of a community, providing a sense of belongingness that has important health ramifications. Greeting is thus a form of cooperation that unites neighbors and prevents them from seeing each other as potentially hostile strangers. In small towns and rural communities, civility can be extended to everyone because the social system is largely closed in the sense that virtually everyone knows everyone else.

Beyond the superficial consideration for strangers implied in greeting practices, there is a universal expectation that people should behave politely in their social interactions. Everyone has some conception of what politeness means, but it is surprisingly difficult to pin down, partly because its definition varies in different societies, and even as a function of social status. A key feature of politeness is avoiding any loss of self-esteem, or any threat to the social status, of others. Anthropologists like to distinguish between positive politeness and negative politeness. Positive politeness means direct expressions of solidarity, agreement, or affection. Negative politeness involves more formal statements or actions designed to protect another's self-esteem or status. Examples include formal apologies, deference to another's knowledge, and avoiding actions or words that might be construed as critical or insulting. Anthropologists conclude that people of higher social status employ negative politeness more often whereas those lower in the hierarchy use more positive politeness. Women's politeness tends to be more positive than that of men.[1]

All societies require, or at least expect, adherence to rules of politeness, which vary depending on the context and society. Thus the decorum about not speaking out of turn is extremely rigid in the US Senate, but not in the bars that some senators frequent following their debates. Japanese rules of decorum are also very different from those of the United States. Thus the Japanese consider it impolite to make a request that is liable to be refused because the person who is asked is thus forced to be unpleasant. Direct statement of opposition is also disapproved of, and the customary way of expressing disagreement is to change the subject. Emotion researchers have noted that the Japanese facial expression of disgust is greatly muted compared to that of Americans. The Japanese are also more disapproving of uninhibited laughter and of teasing that could involve a loss of face for the butt of the joke.[2]

Politeness might often seem superficial but it is actually very important for the smooth negotiation of social interactions, and for accomplishing complex group tasks. Thus, air crews with good safety records show a high level of deference to the captain.[3] Politeness also requires considerable verbal skill. Thus the human capacity for language is surely our most conspicuous adaptation for group living. We do not really know why our ancestors acquired the specialized larynx and brain mechanism regulating fine gradations of sound production and language comprehension. Fossil evidence suggests that the specialized larynx is fairly recent, emerging within the last 300,000 years. Effective group function is certainly possible without a complex language system. Social predators, like African hunting dogs, hyenas, and wolves (see fig. 9), have extremely cohesive societies, with division of labor (between foraging and protecting young, for example), but accomplish this using a small repertoire of signals, including elaborate greeting ceremonies.

Language is essential as a system of shared signals for tying large human groups together. The broad territories of the Roman Empire were united by a shared system of laws that was preserved in writing. Yet, these applications of the language adaptation are emergent

Fig. 9. Wolves cooperate to bring down prey much larger than themselves, as depicted in this Currier and Ives print. (Library of Congress)

properties—they have no immediate relevance for the simpler societies in which language capacity evolved. Once people developed a capacity for speech, they produced complex verbal communication systems, languages, that existed independently of any speaker and could be passed on to future generations, once the speech sounds are transposed into written symbols.

Prior to written language, and the formalization that it permitted, language was very much a local matter. Sounds and their meanings were defined by the community of people who spoke directly to each other. Even today, we can detect minor variations in accent among different regions of a country that are attributable to the direct effect of social interactions. There are a few good animal analogies that could be applied to human language. One is birdsong that requires special adaptations for sound production and is partly acquired through social learning, or imitation. Like young children, young songbirds must learn to communicate, and they even have a phase of subsong that is analogous to preverbal babbling in children. Birdsong also has regional variations, or dialects, that probably reflect the cumulative effect of copying variations or "errors" over innumerable generations. This analogy is interesting but it hardly tells us anything about why people acquired language. There are two reasons for this conclusion. First, birds have likely sung for tens of millions of years. Second, birds sing primarily as a warning to competitors to stay out of their territories. Since females generally do not defend territories, they do not sing. Birdsong is not a language in the sense of having an arbitrary connection between the sound and its meaning (it lacks "semanticity," in the jargon of linguists).[4]

Social mammals, including dolphins and whales, as well as elephants, may use auditory signals to coordinate their movements over long distances. Whether these signals should be considered a language is a controversial issue. A better case can be made for honeybee foragers that use a waggle dance to inform hive mates about the location of nectar-bearing flowers. The dance is meaningful. Thus, the vertical axis of the dance tells observer bees about the direction of the food source. It is semantic in the sense of containing meaning, although the connection between the symbol and the meaning is not arbitrary, as in human language where a given object can be represented by different sounds.[5]

Early languages may have united loose assemblages of genetically related hunter-gatherer groups. This would have facilitated important cross-group functions, such as the arrangement of marriages. Yet, it seems likely that the benefits of language function, whatever they were, were primarily in the day-to-day interactions within groups, and within family units, where most communication takes place. Furthermore, it seems clear that language helps members of a social group to coordinate their activities, making it easier for individuals to conform to the needs and expectations of the subsistence unit.

Social conformity is a kind of reciprocal altruism in which individuals help each other by behaving in a predictable way. When they follow the rules of the road, drivers are bound by a set of conventions, including which side of the road to drive on, and how to signal changes of direction, as when turning off a road in front of another driver. Both the giver and the receiver of traffic signals benefit from increased predictability leading to enhanced safety.

All types of etiquette, including acceptable styles of dress, increase the predictability of individual behavior and can be interpreted as a form of reciprocal altruism on this account. In societies with rigid sex roles, women often dress differently from men, for example, and in societies with rigid class systems, social standing often determines styles of dress. By dressing in conformity with one's social role, a person greatly increases the predictability of social interactions he might have with another person.[6]

Conformity is an important element of effective group function for armies, businesses, and all other organized social groups that work for shared goals. Social psychologists have investigated the evil aspects of social conformity, including poor decision-making and genocidal attacks on "out-groups," such as in the Nazi Holocaust against Jews. Conformity can also be bad for individuals, stifling their spontaneity and creativity. In some societies, conformity pressures drive people to overwork, even to the point of dying prematurely, as in the Japanese phenomenon of *karoshi* (or death from overwork).

THE DARKER SIDE OF GROUPS

Humans are designed for life in small kin groups. We are capable of high levels of altruism toward those we identify as friends. When our groups are threatened by competing entities, however, we circle the wagons and visit destructive aggression on other groups. The us-versus-them division runs through our social lives like the light and dark bands in the grain of wood. From our household to our company, our organization, our team, our state, and our country, we are constantly alert for external threats and willing to take action to defend our collective interests. That is why soldiers are willing to sacrifice their lives in attempts to defend their country and why the same act can be viewed as heroism by the in-group on whose behalf it is carried out and as a vicious atrocity by the enemy's group.

Human groups definitely have a dark side, and it is this side that has fascinated American social psychologists. Many, like the Oliners, had personal ties to victims of the Nazi Holocaust. They were strongly motivated to understand why groups like the Nazis can behave in such evil and inhumane ways. The first thought was that the evil is contained in individuals. You have an evil authoritarian leader like Adolf Hitler, under whom operates another tier of evil authoritarians like Hermann Goering, head of the Gestapo, or the German secret police who organized the Holocaust, who were themselves served by vicious officers like Josef Mengele, who used his position as chief medical officer at Auschwitz to conduct atrocious experiments on camp inmates.

Efforts to attribute the cause of Nazi atrocities to evil authoritarian personalities quickly unraveled, however. There was no compelling evidence that people can be neatly divided into the kind of evil people who would play a role in a government-sponsored atrocity and others who would refuse to participate.

The war crimes tribunal at Nuremberg (see fig. 10), which was designed to hold surviving Nazi leaders accountable for their role in the Holocaust, also challenged the notion that the Nazis were hideously evil as individuals. The Nazi leaders were disappointingly ordinary, leading many scholars to speak of the "banality of evil," a

Fig. 10. The Nuremberg Trials: Hermann Goering (standing) is accused of orchestrating the Holocaust while head of the Gestapo, the German secret police. (Library of Congress)

phrase subsequently coined by Hannah Arendt in connection with Adolf Eichmann, who was tried in Jerusalem for his role in the Holocaust.[7] If evil people are very ordinary, could ordinary people caught up in a bad situation behave in very evil ways?

THE ABCS OF GROUPISHNESS

Social groupings apparently preserve their integrity and morale in part by excluding others. Groucho Marx was lampooning this aspect of social clubs when he said, "I would never join a group that would have me for a member." Whether it is a social club or a sports team affiliation, a religion or a state, group memberships tend to be emotionally intense, even though they are also quite arbitrary or accidental. It is well known, for example, that religious affiliations are particularly intense, even though they are largely a function of the accidental geography of one's birth, as even a cursory glance at a map of world religions attests. By and large, Africans are animists, Europeans are Christians, and Middle Easterners are Muslim, for example.

Devout members of religious faiths can often be distinguished by their behavioral habits. Catholics genuflect when they pass churches, for example, and Muslims pray five times a day on a special prayer mat while facing Mecca. Sometimes religious rules of conduct can be personally costly and appear to work against Darwinian fitness. To take an obvious example, many world religions condemn adultery. Others ban intercourse during ritual periods, such as Lent. Many religions are distinguished by restrictive dietary practices that make it difficult for them to share meals with people of other religions. The Jewish kosher laws are an interesting case in point because they require not only that pork should be avoided but that meats and other foods should be prepared according to complex dietary laws. Materialistic, or Darwinian, interpretations of such laws are sometimes possible, such as the suggestion that avoidance of pork either increased the efficiency of meat production in the Middle East, or protected early Jews from trichinosis, a parasite that lives in pig meat.

According to anthropologist Bill Irons of Northwestern University, religious practices may be explicitly designed to be costly as a way of ensuring that adherents can be trusted. Irons describes religion as a "hard-to-fake sign of commitment." It is hard to fake precisely because religious practices are troublesome, time-consuming, and require substantial sacrifices. Despite the possibility that they are designed to be costly, signs of religious conformity can be used for individual benefit.[8] Thus Irons points out that when Turkmen tribesmen travel, they are unusually careful about asking strangers to show them the direction of Mecca, thereby allowing them to meet fellow Muslims, who might be disposed to provide shelter, food, water, or companionship.

The universality of religions among human populations suggests Darwinian advantages, although it is difficult to say whether these derived from social cohesion or direct material benefits, such as better health, insurance against food scarcity, or increased reproductive success. Research on modern populations finds that religion can have important effects on fertility, so that Canada's Hutterites had more than twice as many children per woman as the rest of the Canadian population, for example.[9] Fertility is also much higher in Muslim countries than Christian ones, but this may have as much to do with national levels of economic development as with religious dogma. People who are actively religious experience higher levels of social integration, have lower divorce rates, are less likely to commit crimes, and generally work harder, all of which would tend to improve their health and economic success.[10] Ironically, although religiously devout women tend to have larger families, they report more infrequent intercourse and derive less pleasure from their sexuality. We cannot be sure what influence religions had on fitness in the remote past, but it seems likely that religious conformity relies on mechanisms of group loyalty that were useful in other contexts.

Synthetic group loyalty can be established in a few minutes simply by dividing individuals into one of two groups, based on chance alone, and providing them with a sense of identity. Social psychologists refer to research carried out in this way as the *minimal intergroup paradigm*. Participants are arbitrarily divided up into groups (for example, by flipping a coin), and each group is given an identifying logo, such as a red button or a green one. Participants then have an opportunity to behave altruistically to each other (e.g., by awarding points that can be cashed in for money). They give more points to people in the same (minimal) group as themselves and fewer points to the members of the other group (the out-group). Such in-group favoritism has been demonstrated by social psychologists working in many different countries.

Contrived laboratory situations thus confirm the intuition that human beings divide the social world into "us" and "them" (in-groups and out-groups). Arbitrary social divisions created in the laboratory are evidently based on psychological adaptations for kin-based favoritism. Psychologists have accumulated additional evidence confirming that group biases are deeply entrenched in the human psyche, which can have tragic consequences, including genocide. Belonging to a group elevates self-esteem. That is why so many people become fanatical supporters of sports teams. When self-esteem is threatened, people also become more discriminatory in their attitudes to out-groups. Thus, difficult economic times in Germany during the rise of the Nazi movement increased resentment against what they perceived as affluent Jews. Jews in all professions lost their jobs.

The connection between humiliation and racial hatred has been demonstrated in experiments. Participants were made to seem clumsy. Following such loss of self-esteem, their racial prejudice increased.[11]

People like to function in antagonistic groups (such as rival groups of sports fans, or rival commercial organizations) because this makes them feel good, particularly if their team wins or their company gains market share. Our brains are evidently adapted to filter social information according to an in-group–out-group distinction just as our perceptual systems automatically divide visual scenes into figure and background.[12]

The fundamental us-versus-them distinction is based on kin selection, as suggested by a variety of lines of evidence. The fact that we should prefer individuals who resemble us in some arbitrary trait, like wearing a green button, is incomprehensible if taken in isolation. Yet it does fit in with a tendency to favor individuals who look like ourselves. Virtually all cohesive groups, from nuns to soldiers to Hell's Angels, use shared elements in their dress to forge a common identity among individuals who are not just unrelated but often complete strangers to each other on the individual level. Favoring others who look like ourselves is one of the most important mechanisms through which our ancestors favored kin. Thus men would have increased their reproductive success if they had favored children who resembled them closely, rather than investing in children who looked so different that they could only have been fathered by another man. This is a weak phenomenon, however, since it clearly does not prevent men from loving adopted children who do not resemble them closely.

The tendency to favor others who look like ourselves goes beyond crude divisions of race and ethnic origin. One of the clearest examples concerns whom we marry. Brides and grooms typically resemble each other in physical appearance, level of physical attractiveness, personality, social status, ethnic group, geographical origin, religion, and politics. This phenomenon is known as *assortative mating* and it may guarantee that people who marry are biologically compatible—but not closely related enough to seriously increase the risk of genetic diseases—and thus likely to be reproductively successful. Biological similarity implies some level of relatedness, of course. This does not imply incest, but it is no accident that most of the societies studied by anthropologists promote marriages between first cousins, even though this is legally outlawed in modern societies because of an increased risk of birth defects. There is so much reciprocal altruism between husbands and wives that the kin network benefits if that altruism is maintained within the extended family.[13]

The tendency to favor in-groups is probably derived from kin-based altruism: favoring others who look like ourselves might have been an evolved mechanism for favoring genetic relatives. This argument is strengthened by the widespread use of kin terms in political and religious organizations. Members of mining unions are called brothers, as are monks in a monastery. Many religious organizations recruit young people and provide them with a network of social relationships that replaces their family of origin. In addition to calling each other by kin terms and refraining from sexual relationships, monks and nuns dress in common garb to emphasize their unity (see fig. 11). By looking alike, they exploit our built-in favoritism toward others who resemble us, even in superficial ways, to build a sense of community.[14] Evolutionists see this mechanism as being part of the process of *phenotype matching* through which kin are favored. Bees provide a good example of this phenomenon because they can recognize close relations by odor and attack unrelated bees entering their hive.[15]

Favoring in-groups is the altruistic ingredient of group psychology. Group competition and aggression are the other side, the source of ethnic cleansing, genocidal wars, and many of the other horrors of recent history. Social psychologists have shown how group identification facilitates hostility, even among groups of children.

SUMMER CAMPS GO TO WAR

Humans think automatically in terms of us and them, and this helps to explain why warfare is prevalent in some preindustrial societies. Many hunter-gatherer societies go to war rarely or never because there is little that is worth fighting over. Hunter-gatherers have little real property. The only thing that is really valuable for them is the land on which they hunt game animals and gather vegetable food. Instead of constantly waging war over hunting territories, they observe a traditional division of hunting lands. This is possible because they have a very low rate of population growth and are not constantly in search of new land to colonize.

The ingredients for tribal warfare were clarified by a pioneering experiment conducted in 1954 on summer camp participants at Robbers Cave State Park in Oklahoma.[16] When the boys arrived in camp, they came up with a group name and painted that name on their hats and T-shirts to solidify group identity. After a week, they discovered that there was a second group in camp and that the Rat-

Fig. 11. Nuns emphasize unity in their dress. (Library of Congress)

tlers would be competing with the Eagles in sporting activities to earn points for a trophy that would go to the winning group.

The experimenters observed instant hatred between the two groups. The Eagles developed a negative stereotype of the Rattlers and vice versa. Each group saw the other as sneaky and treacherous. They maintained that members of their own group were brave and friendly. Allowing the boys an opportunity for social interaction did nothing to soften hostility. They taunted and insulted each other. Antagonisms became so intense that the eleven-year-olds behaved as though the two groups were at war. Flags of rivals were burned. Cabins were raided. A riot broke out in the mess hall where the boys began throwing food at each other.

The experimenters realized that they needed to reduce tensions before someone got hurt. No amount of persuasion or opportunity for getting to know each other better made any difference. Finally, the experimenters hit on the expedient of having the two groups cooperate to solve a problem. They arranged for a truck to break down so that the two groups could cooperate in pulling it up a hill. Following this event, negative perceptions declined and the groups began to like each other. Group conflict can be resolved by forcing the combatants to cooperate in solving a joint problem because this compels them to see each other as part of a single in-group.

Some form of identification with an in-group is an essential characteristic of all social species. Most nonhuman animals recognize group mates through odor, whether they are social insects, naked mole rats, or dogs. Human group identification takes place through various sensory systems, including physical appearance and verbal recognition. Odor is less important, although people prefer romantic partners who are not too closely related to them (as reflected in varying histocompatibility protein complexes of the immune system), and this is detected through body scent.[17] Group identification is arguably a psychological adaptation that helps us to function effectively together. At any rate, the process through which people become bonded to groups is very different from other kinds of learning: we normally avoid sources of unpleasant experiences. Yet people can become deeply committed to groups that cause them pain.

IRRATIONAL INITIATIONS

To some extent, the social groups with which we identify are arbitrary accidents of birth. Thus, a person's religious affiliation is very much a question of where on the globe she is born. One's kin group and ethnic/tribal affiliation are also accidents of birth. The strength of our identification with a particular group is very much a matter of experience, however. Suffering with a group may increase our altruism toward its members.

This irrational aspect of group identification is clearly illustrated by hazing rituals, although initiates generally suffer alone rather than in the company of other pledges. In college fraternities and sororities gruesome tortures are carried out as part of Hell Week initiations, although college authorities have done their best to outlaw such practices. The initiations are conducted in private but are a matter of public concern because of scandals associated with injuries sustained by pledges, occasionally leading to death. Richard Swanson at the University of Southern California died at the Kappa Sigma house while attempting to swallow a quarter-pound slab of raw liver, for example. Abandoned in the wilderness on a winter's night wearing inadequate clothing, Frederick Bronner, a California junior, fell into a ravine resulting in serious injuries. Unable to go on, he stayed where he was until he died from exposure. National Greek organizations repeatedly promise to eliminate the most dangerous of hazing practices, but painful initiations are unlikely to disappear—unless the Greek societies themselves are discontinued.[18]

Hazing is deeply irrational. Imagine that you had just visited a new restaurant. You were seated at the table waiting to order. Instead of greeting you, the waiter insulted your appearance. It seems unlikely that you would want to go back. Yet, the pledges who undergo the arbitrary unpleasantness of Hell Week generally do not abandon the organization that has tortured them. Instead, they become lifelong devotees, forming stable associations based on reciprocal altruism.

Social psychologist Robert Cialdini at Arizona State University sees a direct parallel between Greek society hazing and the initiation ceremonies endured by warriors in tribal societies.[19] The different trials endured by pledges are similar to those endured by an African Thonga boy during his stay in the "yard of mysteries," from which he emerges as a man. Thonga boys endure the following trials: beatings, exposure to cold, thirst, eating disgusting foods, brutal punishment, and threats of death.

Moralists like to discover the causes of social problems in individual depravity. This approach does not have much going for it as an explanation of inhumane Greek society initiations. Individuals who carry out hazing rituals are not particularly sadistic or unusual. The average Greek society member is, if anything, more psychologically stable than the average college student based on personality tests.[20] Hazing is not just for the amusement of a few sadistic creeps who take advantage of Hell Week to express their depraved inclinations. Unpleasant initiations are evidently carried out to strengthen group identification and solidarity.

Oddly, people who suffer together may form exceptionally close social bonds as a consequence. This could explain why combat veterans become so closely attached that they often think of each other as family and may go to extraordinary lengths to help each other out, even to the extreme of throwing themselves on a grenade. The social bonding produced by shared trauma might also help to explain the strange phenomenon of kidnap victims, such as Patty Hearst, becoming emotionally attached to their captors. Following her abduction by the Symbionese Liberation Army, the heiress made common cause with the violent, albeit confused, objectives of the terrorists. The group funded itself by robbing banks. Hearst was shown on a bank's security cameras, "voluntarily" participating in an armed stickup. Traumatized by her experiences in captivity, the brainwashed heiress had become "Tania," an armed revolutionary who posed for poster pictures in front of the Symbionese Liberation Army flag. Such identification with captors is often referred to as the Stockholm Syndrome after four Swedish bank employees who were held hostage in a vault for six days in 1973 and became emotionally bonded with their abusers, as reflected in subsequent willingness to work for their legal defense.[21]

Also consistent with the bonding effect of shared trauma is the fact that among tribal populations, those with the most severe initiation rituals also have the strongest group solidarity, according to a study of fifty-four societies. Experiments by social psychologists on painful initiations help to explain why. In one study by Elliot Aronson and Judson Mills, women had to read some sexually explicit material aloud to a male experimenter in order to gain entry to a dull discussion group. They evaluated the discussions as more worthwhile than women who had not been humiliated before joining.[22] Enduring physical pain, rather than embarrassment, had the same effect. The more electric shock a woman received during the initiation—a procedure that would be prohibited today for ethical reasons—the more she felt the discussions were interesting and valuable. These bizarre experiments shed much light on hazing rituals in Greek organizations, and painful initiations in general. It is difficult to say whether this phenomenon is peculiar to large groups or extends to pairs of people. To cite one parallel, women become more intensely bonded to their infants following a particularly stressful childbirth.

Whether one identifies strongly with a social group or not, merely being in the presence of other people changes our behavior. Generally, we try to minimize conflict with other members of a group. Our groupish tendencies include conformity in dress, behavior, and even ideas. Social psychologists have been deeply impressed by the apparent irrationality of human conformity—at least as seen through the perspective of a Western tradition that values, and therefore exaggerates the importance of, individualism. They saw in this phenomenon an adaptation for cooperative functioning of hunter-gatherer groups, long before the term evolutionary psychology was widely used.

Social conformity helps groups to function effectively by greatly increasing the predictability of individual behavior. In the modern world, residents of a given country agree to drive always on the same side of the road. This simple act of conformity in itself greatly reduces head-on collisions and fatalities on the road, but it is accompanied by a much more elaborate set of rules of correct driving. Admittedly these rules are enforced by traffic policing, but many individuals are careful to obey them even if they never receive a fine. Most reasonable people accept that orderly conduct is essential for the safety of the individual driver and others traveling the same roads. Even where much less seems to be at stake, human social behavior is characterized by a surprising degree of conformity. Such like-mindedness facilitates altruism within groups. Unfortunately, it is also the basis of atrocities committed against out-groups.

GROUPISHNESS AND CONFORMITY

Social groups can be formed quite arbitrarily and immediately. Yet, groups that are joined following grueling initiations are more cohesive, their members more willing to make sacrifices for each other. These phenomena are some of the clearest demonstrations of the automatic, or mindless, aspects of groupishness. There are many other examples that fall into the same category. One of the most striking is the high level of conformity of dress and manners that are characteristic of all stable groups, from monasteries to football teams to groups of soldiers.

Groupishness is reflected in common clothing. Uniformity of dress is particularly striking in earlier generations where people dressed formally, often in black-and-white garments. Dress styles today are more influenced by differences between age groups, by the distinction between leisure time and work, and by the intrusion of casual clothing into formal occasions.

Age differences in dress are most obvious in the contrast between teenagers and older adults. Teenagers go through a predictable phase of rebellion that is expressed in clothing styles and habits, like extensive body piercing and visible underwear, that are somewhat offensive to many adults. Adolescent rebellion seems like an opportunity to escape group control over dress, but it is not. While adolescents are rebelling against adult social norms, most rebel in the same way. That is the impression one gets from group pictures of teenagers in different decades. Their "unconventional" clothing is so predictable for any period that it is essentially a uniform. Students adopted denim jeans in the sixties as a gesture of solidarity with the farm laborers who had worn them, for example. This was intended to shock their uptight social-climbing parents. While jeans have been retained as the uniform of college students—and are worn by people of all ages—the emphasis is now on designer jeans, indicating that they have shifted back to preoccupation with social comparison and materialism.

Uniformity in dress is largely self-inflicted. In most situations, one is not told what to wear. Groupishness, the powerful need to fit in with the expectations of others, drives conformity in dress, as well as behavior and ideas. Social psychologists have conducted much research on groupishness, focusing particularly on conformity to group judgments and obedience to authority.

Mental conformity, rather than conformity of dress, first piqued the interest of researchers in social psychology. One important pioneer was Turkish scholar Muzafer Sherif. His interest in group behavior began with the grisly experience of being almost bayoneted by an invading Greek soldier when he was a teenager. During the 1930s, Sherif conducted his conformity research by ingeniously using a visual illusion. He exploited the fact that a stationary beam of light striking the wall of a dark room appears to move. This visual illusion is referred to as the autokinetic effect.[23]

Sherif told participants in his experiments that the research was concerned with visual perception. Seated in a dark room, fifteen feet from an unmoving circle of light, each participant gave an individual estimate of how far the light appeared to move. Judgments were all over the map: one subject might say the light had moved one inch, another that it had moved eight inches.

Participants were then exposed to social pressure by making their estimates in groups of three, on three different days. Sherif found that the difference in estimations shrank steadily from one day to the next. By the third day, all three people gave exactly the same estimate of how far the light had moved. This result suggests that when people operate in groups, they lose their independence of judgment and become essentially mindless conformists. A striking contemporary example of mindless conformity was the European Nazi movement, which provided adherents with racist objectives, invalid beliefs about the Jews and other ethnic groups, and shared patterns of behavior, such as goose-stepping in parades and shouting, "Heil Hitler!" in greeting.

The mindless conformity interpretation of Sherif's results struck social psychologist Solomon Asch as unduly pessimistic. Asch felt that Sherif's experiment was too far removed from everyday experience and that conformity would be less likely if people were given a less ambiguous test.[24] In his experiments, people were asked to decide which of three lines was equal in length to a standard line, which is rather like a carpenter matching up pieces of equal length in constructing a frame. People are good at estimating the length of a line accurately, so everyone could solve this problem correctly. Asch now tested the impact of social conformity pressures by investigating what would happen if several people in a group had already agreed on a wrong line before an individual was tested.

Imagine that you are a participant in such an experiment. You are seated at a table where six other people are assembled. The experimenter presents his cover story, claiming that he is interested in line length estimations. He shows you a chart with a single bold line and another chart with three sample lines. Which of the three lines is the same length as the standard line? The experimenter goes around the table obtaining an answer from each person in turn.

The test is very easy, and there is complete agreement in the responses. The experimenter repeats the process with another set of lines and the outcome is the same. On the third trial, however, the first subject selects the wrong line. What is happening? Has he experienced a slip of the tongue? Incredibly, all of the other four people agree with the wrong line. Have they become visually impaired? Or completely lost their sanity? In reality, the others are confederates of the experimenter, whose answers follow a carefully rehearsed script. Only one person at the table is really being tested.

The confederates selected the wrong line on twelve of eighteen trials. The question is, how would you respond to the group pressure? Would you go along with the incorrect majority some of the time, all of the time, or never? Half of the experimental participants agreed with the clearly wrong answer at least six times (of twelve tests). Only a quarter of the subjects never conformed on any trial.

Conformity pressures are thus strong enough to make most people defy objective realities and agree with the opinions of a group. Yet, such conformity is not mindlessly automatic. Even as they agreed with the majority, participants were visibly upset. They knew that the majority was wrong. Their outward behavior was shaped by conformity pressures but their private opinion did not change. Asch's study thus suggests that people are good at suppressing their private opinions to maintain agreement and harmony within the group. Groupishness is not necessarily inconsistent with protecting the interests of the individual. We may cooperate publicly in groups while nevertheless privately protecting our own interests.

Interesting as Asch's research was for the light it sheds on group behavior, it cannot really be invoked to explain the worst excesses of groups like the Nazis. There is a world of difference between saying the right thing in a social situation and carrying out an order to kill an innocent human being who has done nothing to offend us.

ATROCITIES AS OBEDIENCE

Social psychology was deeply affected by the Holocaust. Many social psychologists were, and are, Jews, who wanted to ensure that a Holocaust would never happen again. The first step was to try to understand why individual Nazis had agreed to participate in atrocities committed against the Jews and other hated minorities, such as Gypsies. There are two broad possibilities. The first is that the Nazis were exceptionally evil people. The second is that although quite ordinary, as individuals, the Nazis felt trapped by their situation and were compelled by social pressures to obey orders that they privately disapproved of. The second explanation has never been a popular one. Movies set in World War II rarely depict S.S. officers as tortured by self-doubt. They are invariably arrogant, self-assured, and sadistic.

Real-life S.S. officers were very different from the stiff-necked Nazi stereotype, played to perfection by Erich von Stroheim in silent movies. Or at least they appeared very different in defeat at the Nuremberg war crimes tribunal. The war criminals seemed shockingly ordinary. Nothing in their faces or comportment revealed the heinousness of their actions. Commentators chose to focus on the evil rather than the ordinariness, however. When invited to account for their appalling actions in furthering Hitler's racial cleansing agenda, the German officers tried at Nuremberg had little to say in their own defense, except the commonly heard claim that they were simply carrying out the commands of people higher up than them and that to do otherwise would have risked death for insubordination. This defense did not play well at Nuremberg because it would have exonerated most of those tried. Adolf Eichmann was the head of the S.S., which was responsible for organizing the incarceration and slaughter of Jews. Even he continued to protest his innocence and to claim that he had only carried out his orders. Eichmann was symbolically tried in Jerusalem, in the young state of Israel, and it is unlikely that any defense would have helped him. Observers interpreted his plea of obedience as patently self-serving. Many Jews were heartily offended by it. They felt that the least they were entitled to was some contrition. Eichmann's defense was a notorious flop in the courtroom also: he was hanged.

Social psychologists were not so sure. Stanley Milgram recast the Nuremberg trials as a testable hypothesis. When people participate in atrocities, they do so either because they are evil people, or because their actions are controlled by the social situation. The most salient aspect of the social situation for German officers was the hierarchical command structure in which they served and the obedience to authority that was demanded of them. We cannot retry Nuremberg and we cannot bring Adolf Eichmann back from the dead, but we can study obedience in ordinary people. Milgram wondered whether ordinary folk are capable of carrying out an atrocity if instructed to do so by an authority figure. He devised an ingenious experiment to find out. Ordinary people are not S.S. officers, of course, but if people like you and me would commit an atrocity when told to do so by an authority figure, then it should be easy to accept that Nazi officers might do the same following extensive military training to carry out orders without question.

In Milgram's obedience experiments,[25] the cover story—used to conceal the true purpose of the experiment—was that the participant was teaching another person associations between pairs of words. These word associations were quite arbitrary and pointless. If the teacher said, "Cat," the learner might respond, "Cabbage," for example, and if the teacher said, "Rope," the correct response might be "Pail."

The person playing the role of learner made many "mistakes." For the first error, the teacher was instructed to administer an electric shock at the level of fifteen volts. For each subsequent error, the shock was raised fifteen volts. Milgram wondered how high people would be willing to raise the shock level when told to do so by the experimenter. Would they raise the shock above the life-threatening level of three hundred volts?

Pulling off an atrocity in the psychology laboratory requires a flair for theater. Milgram designed an impressive-looking bank of electronic devices that produced a series of loud clicking noises followed by an ominous hiss to simulate the delivery of electric shocks. In reality, the machine only made noises and did not generate shocks. The illusion of shocking another person was enhanced by a skilled actor who played the learner receiving shocks. The actor reacted to simulated shocks by grunting in pain and asking to get out of the experiment. Groaning and pleading escalated until the 330-volt level, at which point the actor fell eerily silent, evidently passed out.

Most of the participants were very upset by what they were doing. Milgram characterizes a typical reaction on the part of an "instructor" as follows:

> I observed a mature and initially poised businessman enter the laboratory smiling and confident. Within 20 minutes, he was reduced to a nervous stuttering wreck, who was rapidly approaching a point of nervous collapse. He constantly pulled on his earlobe and twisted his hands. At one point, he pushed his fist into his forehead and muttered: "Oh God, let's stop it." And yet he continued to respond to every word of his experimenter and obeyed to the end.[26]

Instructions from the experimenter followed a strict protocol. Initially, reluctant teachers were urged on with a mild "Please continue," and eventually they heard the brutal command "You have no other choice, you must go on." Upset or not, two-thirds of the participants continued to press switches even after the learner had stopped responding. This experiment suggests that the majority of ordinary people are willing to commit murder if told to do so by a trusted authority figure—even in the absence of any valid reason for hurting the other person. This rather shocking outcome indicates that some of the Nazi officers may have been close to the truth when they claimed that their participation in atrocities was driven by the powerful compulsion to obey authority figures.

Needless to say, many social psychologists were unwilling to admit that the social situation could so thoroughly overwhelm the moral scruples of individuals. In particular, respected college professors at distinguished institutions like Harvard University do not have a record of committing atrocities. How credible is it that someone would be asked to deliver lethal levels of electric shock to another person in such situations? Repeating the experiments in different settings produced similar results, however.

Milgram's findings are not just a freak of the psychology lab. Any setting in which people are expected to obey superiors creates frightening possibilities for abuse of power. Take hospitals, for example, in which doctors are higher in the authority structure than nurses. In one chilling experiment, 95 percent of nurses were willing to give excessive doses of an unlisted drug to their patients when told to do so by telephone by a "Dr. Smith in Psychiatry" whom they recognized as being on the hospital staff roster but had never met. Their willingness to obey someone they accepted as an authority figure overrode several of their written orders for drug administration. One forbade taking telephone instructions.[27]

Ordinary people are evidently capable of committing atrocities when ordered to do so by an authority figure. That was Milgram's conclusion, one that flew in the face of moralistic interpretations of his day.

Why do people carry out appalling instructions, even when they hate what they are doing? Milgram argued that the tendency to obey leaders is built into us by natural selection. Obedience to a senior figure would have enhanced the prospects for survival and reproduction of individuals in the hunter-gatherer bands of our ancestors, which would rarely have contained more than about fifty individuals. Needless to say, leaders in such societies would be very unlikely to orchestrate large-scale atrocities.

In fact, leaders in hunter-forager groups, like the !Kung in South Africa's Kalahari Desert, are a great deal less autocratic and more benign than absolute rulers in modern nation-states. One thinks of dictators like Adolf Hitler, Josef Stalin, Saddam Hussein, and Pol Pot of Cambodia.

Leaders in hunter-gatherer communities contribute their knowledge and wisdom to decisions such as where to set up camp and where to go hunting. Such societies are egalitarian and headmen are very much servant leaders: they have more responsibilities and obligations than power and privilege. Following the wishes of the leader can be an effective method of avoiding internal conflicts. Literally and figuratively hunter-gatherers benefit from the sense of direction provided by a leader as they ramble about in their home range. This increases the probability of successful exploitation of natural resources through cooperative hunting, foraging, and childcare.[28]

Far from being despotic or self-serving, hunter-gatherer leaders would have been motivated by the good of the group, who, after all, would have composed most of their living close relatives. An evolved tendency toward automatic compliance to requests from such benign authority figures would thus have worked out well for the individual in the evolutionary past. Similar trust invested in an Adolf Hitler, a Josef Stalin, or a Saddam Hussein can have frightful consequences for the obedient, their families, and their countries. Such rulers rarely commit their atrocities in person. Instead, their evil deeds are mostly carried out by ordinary people close to the bottom of a tight chain of command. Ordinary people commit atrocities when given orders that they feel they cannot refuse. Some of the compulsion to obey may stem from fear of the bad consequences of refusing orders, but much of it is due simply to a powerful evolved tendency to obey authority figures like parents or tribal leaders. Obedience to authority is one useful adaptation for group living. Another is the division of labor.

DIVISION OF LABOR

Conformity and obedience are hallmarks of human groupishness and help to account for the constructive achievements of modern societies, as well as their atrocities. Behaving in predictable ways is typical of all complex animal societies, of course. Thus, subordinate wolves always greet the alpha pair using a body language that expresses their lowly status. Among naked mole rats, the queen harasses other females, thereby ensuring their infertility and guaranteeing that only she fulfills the reproductive role. Calling her a "queen" makes it quite clear that the biologists who described these remarkable mammals were thinking of an analogy with social insects.

Insect societies are among the most complex in the animal world. Some termite colonies, for example, contain several million individuals, inviting comparison with modern human cities. Such complex societies work best if there is division of labor, not just in terms of reproduction, but also in terms of productive work. Among social insects, larger individuals are likely to be found guarding the colony. Others specialize in cleaning out the colony and taking care of larvae. Some work at construction. Others sally forth in search of food. Such specializations can last throughout the adult life of an individual. Others can be part of a life cycle. In general, younger individuals spend more time working inside the relative safety of the colony, and older ones go outside for the riskier jobs of soldier and forager.[29]

It is rather fascinating that bees, ants, social wasps, and termites have such a complex division of labor. Although much remains to be learned about how these specializations are fine-tuned, the individual actions performed by a worker are largely genetically determined. One of the important hygienic functions of honeybees is to detect larvae that have died, uncap the cells they occupy, and clean out the remains. Larvae are killed by bacterial diseases, such as American foul brood, and removal of dead larvae helps prevent the disease from spreading. Genetic studies have shown that uncapping and cleaning out cells are inherited separately as single gene effects, so that some bees uncap cells but fail to clean them out but others cannot uncap cells but will clean them out after they have been opened by researchers.[30]

A bee colony's collective activities is largely determined by an interaction between the genetics of the individual and the environment inside and immediately surrounding the hive. Yet, individuals are capable of a simple kind of learning, such as remembering the location of a food source. Not only do foragers learn where there is a plentiful supply of nectar-producing flowers, but they communicate this vital information to others. Among army ants, experienced scouts search for colonies of a different species from whom they will steal young workers (i.e., larvae) to serve as their domestic servants, or slaves. During a slave-making raid, the scout lays down an odor trail that the other raiders can follow.[31]

Human divisions of labor are in some ways less impressive than this, at least if you compare the simpler societies of our hunter-gatherer ancestors. The most remarkable division of labor of hunter-gatherer societies was between the sexes. Females spent more of their time working close to their home camp, foraging, caring for children, fashioning clothing and utensils. Men traveled farther from home in pursuit of game animals. This division of labor exists in all subsistence societies and may affect occupational choice in the modern world, even in sexually egalitarian societies like the United States, in which the great majority of elementary-school teachers are women and the great majority of engineers are men. Interestingly, there is a fairly strict segregation of work roles according to sex in all societies studied by anthropologists, but the jobs of men and women are not entirely consistent. In one society, all of the food preparation and cooking might be done by women, but in another, all of the cooking is done by men, for example. In modern societies, men and women perform similar functions in similar jobs and occupations, although representation of the sexes in various fields of employment is not equal.[32]

It seems obvious that men have brain specializations for hunting. For example, they are much better than women at hitting a target with a projectile and have substantially faster reaction times. It also seems clear that women are biologically predisposed to be nurturing: they not only do most of this unpaid work, but are drawn to it from an early age, an attraction that is reduced by medical conditions, such as the adrenogenital syndrome, that masculinize the female brain before birth by exposing it to too much sex hormones.[33]

In the complex modern world of cities people have an astonishing ability to specialize in various occupational roles that have little to do with hunting and gathering. We not only perfect artificial job roles that allow us to make a living. In addition, we learn a great deal about the occupational specializations of others. This kind of social intelligence is useful for the smooth functioning of a complex society. For example, if we are confronted by an armed thug, we are likely to behave very differently than if stopped by an armed police officer because we expect cops and robbers to behave differently. Social psychologists conclude that ordinary people can assume many complex roles without training, possibly reflecting our evolved groupish tendency to fit in in social situations. Some of these roles might spur us to heroism but they can also lead to atrocities, as demonstrated in a notorious Stanford University simulation of prison life.

ATROCITIES IN PRISON

Virtually any human social interaction relies heavily on implicit knowledge of underlying social roles. To play the role of a patient effectively, one must know the role of the doctor, and vice versa. We thus have knowledge of roles that we may never have the opportunity to play. Some of this knowledge is derived from personal experience, but much is also cobbled together from fictional accounts. Knowledge of occupational roles may be patchy, particularly if the job is specialized and complex. Thus, it would be nearly impossible for an untrained person to successfully impersonate a brain surgeon, even for a single operation. On the other hand, it would be much easier to impersonate those whose jobs we observe frequently, such as waiters, police officers, and clergy.

Most people have never visited a person in prison, yet we carry around with us a rather detailed script for how prison wardens and prisoners are supposed to behave, judging from a notorious prison simulation experiment.[34] Presumably we build up these scenarios from a combination of journalistic snippets and fictional depictions in movies and novels. However they are acquired, social roles can exert a powerful control over our actions so that we behave in predictable ways, thus engaging with the social roles of others rather like the engaging teeth in gear wheels. Such predictability is a form of groupish altruism that favors joint action and social cohesion.

According to the findings of a prison-guard experiment by Phillip Zimbardo, social roles can also be important in accounting for atrocities committed against out-groups even when no explicit orders are given. Many atrocities occur spontaneously as individuals play out their social roles.

Zimbardo, a social psychologist at Stanford University, was intrigued by some of the peculiar features of modern prison life. These overcrowded institutions are alienating, degrading, and oppressive. There is palpable tension, although prisons are less dangerous than is commonly supposed. Nevertheless, prison guards sometimes behaved brutally. Prisoners became depressed, exhibiting symptoms of passiveness and demoralization. Zimbardo wondered whether these problems were due to the social situation or were more a feature of the kind of people who end up in prisons, whether as convicted felons, or corrections officers. What would ordinary people do if they were artificially assigned to roles as prisoners and guards?

Two randomly formed groups of students played the role of either a prisoner or a guard in a simulation of prison life conducted in the basement of Stanford University. Few instructions were given and the students apparently filled in the gaps from their own knowledge of the roles of prisoners and guards.[35]

Some aspects of the experience of a typical prisoner were faithfully replicated. Prisoners were arrested unexpectedly at their homes, thanks to the cooperation of local police, and put through the usual rigmarole of being booked, fingerprinted, and admitted to prison. Inmates were easily identifiable by their loose-fitting smocks and ankle chains. Khaki guard uniforms were complemented by sunglasses, night sticks, whistles, and keys.

Prisoners were entitled to three bland meals per day and three supervised visits to the toilet. Individual prisoners were identified and endured the usual prison routine of being lined up and counted. Participants were on their own otherwise. The researchers sat back to see what would happen.

The outcome was disturbing. Guards let power go to their heads. They shouted at the inmates and subjected them to inhuman and degrading punishments. Some prisoners were forced into crowded cells, for example. Others endured solitary confinement. Prisoners were pointlessly disturbed in the middle of the night. They were lectured, intimidated, and forced to do push-ups to the point of exhaustion.

Prisoners initially objected to their cruel treatment but this only brought retaliatory abuse and they gave up their resistance. The first prisoner had to be released after only three days. He was acutely depressed. After six days, the experimenters called a halt for ethical reasons. Letting the experiment continue for two weeks would have exacted too heavy an emotional price.

How do we explain the sudden emergence of brutal and dictatorial habits in a group of liberal college students during the heyday of the hippie movement? According to Zimbardo, the students' assigned roles had exerted a powerful influence on their behavior—so powerful that there was risk of real harm being done to the captives. Prisoners were exhilarated at their release. The guards, by contrast, had been enjoying themselves and were disappointed by the premature ending of the experiment.

Zimbardo's experiment demonstrates how individuals can accommodate themselves to social roles that are very different from their ordinary lives. Prison guards do not have to be brutal, of course, but untrained individuals who are suddenly given the prerogatives of a corrections officer, without the training and oversight prison guards normally receive, are liable to abuse that authority. Zimbardo organized a meeting of the former guards and the former inmates to discuss what they had learned from their experiences. Many of the prisoners complained bitterly about the abuse they had endured from the guards. All agreed that the experiment had taught them a great deal about the human potential for brutality that can be unlocked by a social situation and the social roles we play in that situation.

Like Milgram, Zimbardo concluded that moral concepts are not particularly helpful in accounting for the atrocities perpetrated on one group by another. Their work is just one part of a large body of research indicating that human altruism, and its opposites, aggression and discrimination, is very much a feature of the social environment, rather than the individual. Perfectly ordinary people are demonstrably capable of extremely disturbing behavior when affected by powerful social influences based on compliance with authority or conformity to social roles. Understanding the causes of atrocities does not in any way mitigate their harmfulness, of course, nor does it undermine the legal tradition of holding each individual responsible for his or her actions. Thus, rescuers of Jews during the Holocaust refused to give in to conformity pressures.

Wherever we find them, atrocities against out-groups can be considered an unfortunate consequence of evolved groupish tendencies that help in-groups to function effectively. This is not to say that groupishness evolved because it favored groups at the expense of individuals. It is more reasonable to assume that the inclusive fitness of individuals was improved if they were good at cooperating with other members of their subsistence group (the majority of whom would have been blood relatives).

Being adapted for group life explains the remarkable capacity of human beings for altruism toward strangers, as well as hostility toward antagonist groups. Apart from facilitating atrocities, groupishness has many other disturbing features, including the mindless conformity of religious cults and political fanatics, the intellectual numbness of group decision-making, and the mediocrity and laziness of employees in many large corporations. Failing to pull one's weight, or "social loafing," as social psychologists refer to it, is an almost inevitable consequence of group efforts in which the precise contribution of any individual is difficult to measure.

Despite the expectation that people in large complex organizations should do very little work while getting the most out of their jobs in terms of pay and privileges, some individuals do the opposite, putting jobs before families and work responsibilities before happiness and health. Conformity is not just dangerous to out-groups, for people who work too hard, it can be dangerous to the conforming individual as well.

DEATH FROM OVERWORK (*KAROSHI*)

Work can be very dangerous. In some fishing communities, men are washed off their boats and die of drowning or exposure almost every year. Being a soldier is the most dangerous, and hence the most altruistic, of occupations in which individuals voluntarily put their lives at grave risk in order to defend their families, communities, and way of life. Even such extreme risks can be understood in evolutionary terms because they are balanced by corresponding benefits. Fishermen who are too fearful will never feed their families. Societies that cannot mount opposition to invaders will be quickly overrun. What are we to make of people who are otherwise healthy but drive themselves to an early grave through overwork?

This phenomenon, known as *karoshi*, first came to public attention in Japan during the economic boom of the 1980s. In many highly publicized accounts, the victim, usually a man in his forties or fifties, died suddenly, typically of a stroke, brain aneurism, or heart attack, following many weeks of working long hours. The medical antecedents were fairly well understood in terms of stressful overwork. They had increasing blood pressure, leading to hardening of the arteries and other kinds of cardiovascular damage that led to collapse and death. Since then, it is estimated that there could be as many as ten thousand victims per year, although official statistics number only in the hundreds.[36] These discrepancies have to do with how *karoshi* is defined. The official statistics are restricted to individuals actually dying on the job following extensive overtime. More liberal definitions include all those who acquire life-threatening health problems as a result of prolonged overwork.

The problem is so widespread in Japan that a special pension fund has been established for *karoshi* victims' families. Dramatic early deaths on the job may be only the tip of the iceberg as far as the dangers of work are concerned. Some people may commit suicide due to the stress of their work lives. The stress of overwork contributes to anxiety and depression. People who work long hours are more vulnerable to chronically high blood pressure and heart disease. They have little time to exercise, so that they may be in poor physical shape, and thus more vulnerable to heart disease, cancers, and metabolic disorders, like secondary diabetes. Having little time to spend at home undermines family life and means that the social support normally derived from a supportive family may not materialize. Instead of protecting the worker's health, family life may become a further source of stress and illness.

How can we explain overwork to the point of undermining health? One common explanation is that it is not voluntary. *Karoshi* may be a feature of Japanese life in part because of poor legal protections for the rights of workers. Thus many companies require their workers to perform unpaid overtime, even though this violates Japanese laws. No records are kept, so that such service overtime is difficult to prove in court. Middle-aged workers may be forced to stay in exploitative companies because it is very difficult for them to change jobs.

Long hours are often expected of managers, who do not fill out time cards and are thus not protected by legal restrictions on work hours. Take the well-known case of Kazumi Kanaka, a sales manager at Toyota Motors and a victim of overwork, who slipped into a coma in 1991.[37] Kanaka worked a seven-day week, usually for more than twelve hours per day, sometimes for sixteen hours. The stress of overwork evidently contributed to gout, which led to meningitis. Despite years of painful attacks, he never got to the hospital for a checkup due to busyness of his work schedule. By using the records of a security company that controlled access to Kanaka's office, his wife demonstrated the long hours her husband had worked and successfully sued Toyota. The company agreed to pay compensation for her husband's death.

A similar case of overwork involved Toshitsugu Yagi, an office worker, whose medical history is more typical of the phenomenon of *karoshi*. Yagi worked such long hours that he often stayed at a hotel near his office, too tired to make the trip home. On some occasions, he worked through the night. Overwork brought promotion but the price was steep: he died of a heart attack at the age of forty-three.

Clearly, office work can be surprisingly dangerous in Japan. Workers are at higher risk of *karoshi* if there is a heavy work load that is aggravated by urgent deadlines. Working in positions like sales, where individual initiative is at a premium and there is little sharing of responsibility with others, is also a risk factor.

Death by overwork is not peculiar to Japan, but it is no accident that it has surfaced as a major problem there. Japan is a cohesive society in which social relationships of family, as well as work, are stable and close. Japanese corporations used to take a paternalistic interest in the welfare of their workers, guaranteeing them lifetime employment and expecting a very high level of effort in return. Hard work was encouraged as a way of promoting the interests of the company, and therefore of its employees. Pressure came from the bottom as well as the top because a slacker let down the entire unit in which he or she worked. Coworkers were often friends who chose to go on family vacations together. Such an environment encourages reciprocal altruism because it evokes the codependence of a subsistence society where the efforts of the individual affect the community.

Today, the relationship between workers and corporations is different. Economic contraction forced many Japanese corporations to downsize their operations. Lifetime work guarantees took a beating as employees were reluctantly fired. This environment of uncertainty has done nothing to reduce the Japanese work ethic and has actually put more pressure on managers to reach corporate goals using limited resources. Hence their long work hours and increased health problems.

Some individuals are more conscientious than others, and this would make them vulnerable to overwork, and *karoshi*. Yet, the social pressures that make people work so hard that they injure their health are specific not just to Japan, but to some classes of Japanese white-collar workers. Americans work long hours, many are workaholics, but the problem of *karoshi* has not emerged as a major health problem in this country, as it is in Japan.

Self-destructive altruism, like *karoshi*, is rare. We are much more likely to hear of the opposite problem, individuals whose altruism fails in a way that can be damaging to others.

CHAPTER 9

When Altruism Fails

People are capable of a high degree of self-sacrificial altruism on behalf of family members, friends, colleagues, and complete strangers. Yet, we can all point to situations in which altruism was expected but did not materialize. The milk of human kindness ran dry. Some parents abuse and even murder their own children contrary to evolved mechanisms of child nurturance. Pedophile priests perpetrate a shocking breach of trust that violates the contract of reciprocal altruism uniting clergy with their congregations. Social psychologists have demonstrated some remarkable instances of people failing to help others in extreme emergencies, such as a person having a heart attack. Mob violence is another example of extreme insensitivity to strangers. Then there is homicidal road rage in which two complete strangers may attempt to run each other off the road over an imagined insult. So much for altruistic conformity to the rules of the road!

The fact that altruism often fails is grist to the mill of pessimists and cynics who are inclined to believe that humans are either depraved or lacking in scruples. The human depravity argument assumes that we are intrinsically drawn to bad actions that harm ourselves and others. This is the original-sin perspective of Christian theology that is eloquently captured by the words of the Catechism rote learned by elementary-school children in Ireland in the 1960s:

> Question: What is the meaning of original sin?
>
> Answer: Our wills are dark, our intellects are weak, and our passions incline us to suffering and death.

The cynical view, that people are always out for number one, is slightly less pathetic but it still condemns us to living in a world in which the natural life of the individual is nasty, brutish, and short, as memorably expressed by political philosopher Thomas Hobbes (1588–1679), who advocated strong government to save people from the bad consequences of their conflicting selfish interests.

People do not always behave as altruistically as we might expect them to, but this does not mean that the pessimists and the cynics are right. On the contrary, evolution has equipped us with the capacity to put the needs of others before our own immediate needs. The mere fact that altruism sometimes fails does not discredit its existence. To argue in this way is rather like claiming that the internal combustion engine does not exist because one's car won't start. Just as engineers have carefully designed functional automobile engines, evolution has fashioned a capacity for human altruism. Understanding what human selflessness was designed to do can help us to understand some instances in which it fails.

THE BAD SAMARITAN

The biblical parable of the Good Samaritan (Luke 10:25–37) is a moving account of an act of compassion between two strangers. This relates to three travelers encountering a man lying half-dead on the roadside after he had been robbed and beaten. The story has many nuances. Thus, the person needing help had just been attacked by robbers and beaten mercilessly. This suggests that there is a lot of Hobbesian nastiness out there. Another interesting wrinkle in the story is that the victim is passed by a priest, who averts his eyes and goes on his way, indicating that those who enjoy moral authority among us can be complete hypocrites. The second passerby, the

Levite, should help because he is a priest and because he comes from the same region as the victim. Finally, help comes from the most unexpected quarter when the Good Samaritan—a man of lowly social standing from a different province—is moved to act by pure compassion for the sufferings of a fellow human. The punchline of the story is that even though there are many differences between us, we can be potentially united by acts of kindness that transform strangers into neighbors.

In modern sensationalist journalism, a favorite genre resembles the Good Samaritan story except that the altruistic passerby is replaced by an aloof onlooker who does nothing as the victim is brutally attacked. For convenience, he or she can be called a Bad Samaritan. Take the notorious case of Kitty Genovese, an unfortunate young woman who was murdered in New York even though her cries for help were heard by at least thirty-eight people. The attack, by a knife-wielding assailant, took place at night in a Woodside, Queens, neighborhood. Kitty Genovese desperately fought off her attacker for at least half an hour, during which time she repeatedly called for help. No one responded, not even to the extent of calling the police. Newspaper editorials had a field day lamenting the callousness of modern life and the depraved indifference to the suffering of others that is allegedly characteristic of urbanites. Social psychologist Bibb Latane of Ohio State University at Columbus took a more charitable position.[1] He suggested that the primary reason for inaction when there are many other people around, as in the Genovese murder, is the assumption that someone else is taking care of the problem. He referred to this as diffusion of responsibility.

The diffusion of responsibility theory makes one central prediction that goes against common sense. The more people there are around in an emergency, the less likely it is that the victim will be helped. Latane tested out his theory in a number of ingenious experiments. In one, smoke billowed into a room through a ventilator near people waiting to participate in an experiment. The more people there were in the room, the longer it took for them to alert the researcher to the problem. If there was just one person, he ran next door as soon as the smoke began coming through the air conditioning vent. If there were twenty people in the room, they waited until the smoke was so thick that they could hardly see each other before reacting. The interpretation is fairly simple. If you are the only person in the room, you know that you are responsible for doing something about the problem, so you react without delay. If there are many other people present, you also feel that something should be done but are inclined to assume that someone else, perhaps the person nearest the smoke, should be doing something. You therefore wait until the smoke is intolerable before doing anything.

Latane's findings indicate that failure to respond to a person in distress is not due to the depraved indifference of urbanites. On the contrary, it is our groupish tendencies that prevent us from reacting. The bystanders were not indifferent to the plight of Kitty Genovese. Instead, the sheer number of people around undermined their responsibility to help. Individuals are more likely to help than groups of bystanders are because being a member of a group, however ephemeral, saps individual responsibility. The failure of individuals to help people in distress is thus often a failure of *groups*. What looks like proof positive of the depravity of modern life turns out to be nothing of the sort. It is hoped that more widespread knowledge of the problem of groupish diffusion of responsibility will help people to take more individual responsibility for what is happening around them. If a crime is being committed, or if there is some other emergency, it is dangerous to assume that someone else is taking responsibility.

Mob behavior is another troubling example of loss of personal responsibility in group situations. Acting in concert, groups of individuals feel free to unleash their antisocial impulses in senseless violence that is directed at whoever or whatever happens to be available. This formula helps explain phenomena as diverse as the street mobs of the French Revolution to the soccer hooligans who attack rival fans, trash stadiums, and pillage stores.[2] Interestingly, when police are confronted by rowdy groups, they are prone to the same loss of individual responsibility, which can result in police brutality. Excessive use of force is preventable if riot police wear name badges, making each personally accountable for his actions. The ubiquity of video cameras also means that no police officer can assume anonymity, as illustrated in the use of graphic videotaped evidence against Los Angeles police in the Rodney King police brutality episode.

Aside from the specific case of altruism being undermined by crowd situations, there are many examples of private actions by individuals that are sometimes interpreted as evidence of intrinsic human depravity. They range from infanticide, child abuse and neglect, pedophile priests, and street crime to road rage and the destruction of public property.

CHILD ABUSE

Children are at greater risk of violence at the hands of nonrelatives, but there are many cases of extreme neglect, physical abuse, and even sexual abuse of children by their own parents. These cases are rather shocking precisely because they violate our expectation that parents should protect and nurture their children.

The first case, breakdown of parental nurturance, is well illustrated by several highly publicized recent cases of mothers killing or attempting to kill all of their children.[3] Such tragedies are attention-grabbing because they shatter our stereotypes of mothers as unfailingly altruistic and nurturant toward their offspring. Though these cases differ in many ways, they have recurrent themes. One is that the mothers sometimes have a history of serious psychological disturbance. Another is that prior to the murderous attack, some appeared to be devoted mothers.

Mothers who kill their children are a rare and extreme case of abusive parenting. From an evolutionary perspective, their actions violate the expectation that parents should strive to increase their reproductive success. Killing older children is particularly destructive of this goal because so much has already been invested in them.

To many, this fitness-calculating approach may seem callous, but the age of the potential offspring is of critical importance. In our own society, killing a fetus at the age of five months is not a crime: it is considered a right of the mother. A few months later, after the child has been born, he or she acquires the legal rights of a citizen, whose murder is a grave crime. Infanticide is practiced, as a form of fertility reduction, in many of the societies studied by anthropologists, however.[4] In some societies, like the Yanomamo of South America, it is common to dispose of one member of a pair of twins, on the assumption that it is almost impossible for a mother to raise two healthy infants of the same age. Infants showing signs of physical imperfection, birth defects, or mental retardation may also be destroyed, allowing mothers to invest their energies in healthy infants.

In such cases, the killing of infants is consistent with the evolutionary design of increasing the number of *viable* offspring raised to maturity, where viable means capable of reproduction, as well as survival. Given the widespread availability of both contraception and abortion in this country, infanticide, or the abandonment of live infants, is rare. When it occurs, mothers are likely to be young, unmarried, and in a poor economic position to raise children alone.

Even if the killing of children by parents is comparatively rare, there are other, more common forms of withholding of parental altruism. They include neglect, emotional abuse, physical abuse, and sexual abuse. These phenomena are puzzling to an evolutionist because child maltreatment of various kinds can have severe physical and psychological consequences that reduce the social success, and reproductive potential, of children and thus the inclusive fitness of abusers.

Examples of some of the ways in which childhood abuse undermines reproductive fitness include the following:

- Shaken babies may sustain severe injuries that permanently impair cognitive function, thus adversely affecting economic and social success.
- Abused children are more likely to suffer from depression and other psychological problems and have a greatly elevated risk of making suicide attempts.[5]
- Emotional, physical, sexual, and psychological abuse all increase vulnerability to a variety of diseases through different mechanisms. The implied lack of closeness in parental relationships can impair growth and suppress immune function

from an early age. Children who experience a lack of affection are at greater risk of behaving in unhealthy ways, are more likely to smoke cigarettes and to abuse alcohol.[6]

- Sexual abuse of girls is associated with early onset of consensual intercourse and early reproduction. Such timing of first births can have adverse consequences for long-term reproductive success. For example, it makes subsequent marriage more difficult, and unmarried women generally have fewer children. Children raised by young single mothers are less likely to achieve social success, which has negative implications for inclusive fitness.[7]
- Abused children are more likely to mature into abusive parents, although most do not.[8]

An entire book could be filled in describing ways that children are harmed by failures of parental altruism, and many such books have been written already. Most assume that abusive parenting is a response to stressful environments. According to this view, parental altruism is difficult under any circumstances but breaks down under the social stress rather like a bad car that labors up most hills but burns out when it is halfway up a high mountain. Psychological stress in the lives of parents is obviously an important factor in many cases of extreme punishment. When parents are worried, their tempers flare up without much provocation and child discipline can cross the line into physical abuse.

Stress is important but it is unlikely to be the whole story. Parents may also be emotionally distant, withdrawing physical affection and failing to provide comfort and emotional support when a child is distressed. Nor can psychological stress explain the widespread use of corporal punishment in warlike preindustrial societies. Corporal punishment is used evidently because it helps to inculcate aggressiveness.

When evolutionists find that the real world does not match theoretical predictions, they may ask whether the anomaly is due to unique features of the modern environment, which differ from the hunter-gatherer ecology to which our subsistence ancestors were adapted by some two million years of evolution. Abusive parenting does not appear to be due to the modern environment for two reasons. The first is that most parents are not abusive as far as we can tell (and these data are notoriously unreliable). If the modern environment were the culprit, we might expect most parents to be abusive. The second reason is that although most parents in existing subsistence societies are indulgent toward their children, they routinely expose them to traumas. One of the most obvious is the painful initiation rites that are frequently considered necessary to pass into adulthood. Nisa, a !Kung woman growing up in a hunter-gatherer society in the Kalahari, recalls a great deal of childhood trauma, much of it revolving around feelings of rejection following weaning and the birth of a younger sibling.[9] There was also a great deal of marital conflict, domestic violence, and even homicide in this supposedly peaceful society. Children would have been exposed to much of this traumatic violence due to the general lack of privacy in !Kung society.

As far as the frequency of child abuse in the United States is concerned, different studies have produced widely divergent results. The problem of inconsistency can be addressed by combining the results of different studies. In national surveys, approximately 10 percent of parents (ranging from 5 percent to 14 percent) admit to physical abuse of their child in the previous year. Approximately 23 percent of adult women recall physical abuse as a child (ranging from 11.5 percent to 34 percent). About 8 percent of people say they were physically abused as adolescents. Most physical abuse occurs outside the family, however, and parents are responsible for only about one-tenth of assaults on children aged ten to sixteen years, according to a national survey.[10]

Reported rates of child sexual abuse vary even more widely. When all of the studies were statistically combined using a meta-analysis, it emerged that approximately one-third of females are sexually abused (30–40 percent) compared to one-seventh of males (13 percent). The estimate for males is less reliable because it is based on fewer studies. Sexual abuse of children by biological parents is comparatively rare. Thus, in one US study, by Heather Swanston and coworkers, 20 percent of abusers were classified as "parent figures," which includes stepparents as well as biological parents.[11] Other abusers were relatives (24 percent), authority figures or caregivers (17 percent), family friends (14 percent), acquaintances (12 percent), and strangers (13 percent). David Fergusson and coworkers found that even fewer abusers among Australians were parents (6.8 percent), and a tiny fraction (1.5 percent) were biological parents.[12] In the Australian study, 6.8 percent of abusers were siblings or stepsiblings, 9.8 percent were other relatives, 18.2 percent were family friends, 6.8 percent were boyfriends or girlfriends, 22.7 percent were other acquaintances, and 28.8 percent were strangers. The evidence thus suggests that biological parents are the *least likely* category of persons to be guilty of sexual abuse of children. This is exactly what would be expected from the high level of altruism shown by parents toward children.

The fact that abuse is higher among stepparents than biological parents can be partly explained in terms of different histories. Biological parents are more likely to have raised a child from infancy and thus cannot see them as potential sex partners. In stepfamilies, a spouse often enters a home in which there are sexually mature female children who are socially, and morally, very inappropriate as lovers, but may be biologically vulnerable because the stepparent has not been present since infancy.

Although a substantial fraction of children experience some form of abuse during childhood, the great majority of the abusers are people other than parents. The fact that parents do sometimes engage in physical or sexual abuse of their children, however rarely, is puzzling. Such unusual instances are obviously cases of the failure of mechanisms of parental nurture. Possible explanations include social stress, disruption of the normal parent-child relationship due to separations, and societal differences in the use of severe corporal punishment. Parents may be rather unlikely to actively abuse their own children, except in extreme situations where social stress erodes the natural affection of the parental relationship, but merely withdrawing support or affection can also be harmful to children. Such reductions of parental investment are unfortunately quite common due to the rise of single parenthood and divorce.

ABSENT FATHERS

Children whose parents divorce are more vulnerable to many of life's serious problems. They are much more likely to suffer from anxiety and depression, to experience alcoholism and drug addictions, to get in trouble with the law, and to have conflictual relationships with intimate partners and children of their own.[13]

These differences between children of intact and divorced families are large and uncontroversial. Interpretation of the differences is another story. Are the adverse life effects of children of divorce due to reduced parental altruism? Or, are they due to genetically inherited personality traits?

If genes were the only influence, then parental divorce as such would be irrelevant. This extreme genetic determinist view has been seriously advocated by scholars in the field but it seems unlikely for several reasons. One is that children who are picked on by their parents are much more likely to become delinquent than their siblings who receive kinder treatment. This suggests that how a person is treated is more important than his genotype in accounting for delinquency.[14]

The other major piece of evidence that genetics is relatively unimportant is that there are huge historical differences, such as the rise of childhood depression in recent decades, that cannot be explained in genetic terms, because genes hardly change at all from one generation to the next. (While it is possible that depressed children are more likely to be noticed, or diagnosed today, this cannot be the whole story: clinical depression is severely incapacitating and thus almost impossible to miss.) Rising rates of depression and social problems can be explained in terms of altered relationships between parents and children, however. Parental divorce is one major source of conflict in modern families that disrupts relationships between children and both of their parents, particularly the one who is absent from the home.

Why would parents divorce if it harms their children? The obvious answer, of course, is that they sacrifice their children's future happiness to find immediate romantic fulfillment for themselves. This glib solution to the problem of why parents would hurt their children, instead of helping them, may be wrong, however. In most cases, the fabric of a marriage is severely frayed before couples decide to separate. By that point, conflicts have come to the surface, spoiling the atmosphere for all in the household. Research suggests that children are most damaged by parental fighting.[15] If a relationship is highly contentious, children may actually be helped by the departure of one of the combatants from the home.

If this theory is correct, then the real crux is not why parents divorce but why they squabble in a way that is harmful to their children. This may not be nearly as much of a voluntary choice as it appears, as can be appreciated by a historical perspective on changes in American divorce. At the beginning of the twentieth century, divorce was very much of a minority problem, affecting approximately one marriage in twenty. One hundred years later, divorce had increased by a factor of ten and it now affects about one-half of all marriages.[16] Clearly, the current social environment makes stable marriage more difficult and unlikely. Some possible reasons underlying the decline of marriage include:

- Increased entry of women to the work force. This increases their economic power and ability to escape an unsatisfying marriage. It also allows them to meet other men who could replace their current spouse, something that would have been more of a problem for the domestic women of earlier generations who were largely confined to their homes. As women's wages, and labor participation, have increased in developed countries, there has been a steady rise in divorce rates.
- Declining fertility may also be important. Couples with several children are very unlikely to divorce for a variety of reasons, including economic practicality, and the fact that children do provide couples with an important motive for staying together. In countries where women still have many children, divorce rates remain low.

There is no doubt that some parents deliberately abandon, or abuse, their children but they are in a small minority, most of whom have some mental illness. No such easy excuse condones the conduct of priests who have used the power and privilege of the cloth to prey upon the sexual innocence of children entrusted to their care.

PEDOPHILE PRIESTS

Sexual abuse of children is most often perpetrated by people who are well known to them and to their families, and may actually take place in the home. A substantial proportion of offenders are well known to parents and frequently enjoy their confidence.[17] Priests clearly fall into this category, or at least they did before the many shocking revelations of the 1990s about priestly misconduct and official complicity in recycling abusive Catholic priests in new parishes gradually seeped into public consciousness, following many scandals and many Machiavellian attempts to shield the church from adverse publicity.[18]

Most psychologists agree that childhood sexual abuse has very damaging psychological effects in later life. The abuse may be traumatic and evoke posttraumatic stress responses, particularly if it is repeated over a long period, for example. Even where victims go on to experience ostensibly normal sex lives, maintaining stable marriages, and raising children, they frequently describe their lives as being ruined by the abuse. These arguments were made convincingly enough in courts of law to yield large damages claims for some victims.

Victims of priestly sexual abuse often point to the breach of trust and sense of betrayal at being assaulted by someone they had learned to trust as a moral exemplar.[19] Many felt helpless to prevent the unwanted sexual advances of their priest, fearing that if they reported the abuse they would be suspected of lying. In some notorious cases, priests described their activities as "confessional" and therefore conducted under the seal of privacy. One villain actually used the confessional itself in order to molest children. Perhaps the biggest source of mental anguish for victims is the sense of being complicit in a sexual relationship that they did not want but felt helpless to avoid. The longer the abuse lasted, the more compromised they became and the more difficult it was to break the seal of secrecy.

Sexual abuse by priests challenges a child's identity in other fundamental ways also. It challenges his religious faith. After all, how could a benevolent deity have such diabolical representatives? Male victims also question their sexual orientation. Our society divides people into gay and straight, and males who have sex with other males are defined as gay.

Losing one's confidence in religion, adult authority figures, and one's sexual identity can evidently be quite devastating. Why would people, like priests, who were held in such high esteem by their congregations engage in such an appalling breach of trust that threatens their occupation by violating the basic obligation of priests to labor for the spiritual well-being of their flock?

The great majority of priests probably do not violate their reciprocal obligation to the community in this way. There are no relevant statistics on the matter but it seems reasonable to assume that most priests do their best to live up to their vows of chastity. Many leading American Catholics now believe that priests should be allowed to marry. The underlying assumption is that chastity is a very difficult ideal, which is potentially damaging to priests and has little value for the church community. On the contrary, according to this view, sexually frustrated priests are more likely to offend against children because they lack a normal outlet for their sexual desire.

How many priests actually do abuse children is impossible to estimate. Researchers with the *Boston Globe* have unearthed evidence of thirty-five serially offending priests who were being rotated to new parishes by their superiors, often with glowing recommendations.[20] If this were the extent of the problem, and it almost certainly is not, then the proportion of confirmed child molesters in the priesthood would probably not be any larger than the proportion in the general population. Given the many opportunities that priests have for unsupervised contact with children, some pedophile priests assault hundreds of victims and may feel immune from prosecution, so that large numbers of children are harmed.

Opponents of celibacy believe that enforced sexual abstinence may make priests more likely to abuse children, not just because they are sexually frustrated but because the priests are themselves in a state of arrested sexual development.

According to human sexuality researcher and writer Vern Bullough, priests are particularly likely to be sexually attracted to adolescents (but not young children) because of the way that sexuality is managed in seminaries. Technically speaking, very few priests are pedophiles but many more are ephebophiles, who are attracted to young adolescents. Some experts believe that priests are more likely to be attracted to young people because of their own arrested sexual development and general emotional immaturity. Bullough believes that the practice of inducting young boys to seminaries cuts them off from opportunities for normal sexual development. Seminary students are given very little practical sex education and are not fully informed about the difficulties of maintaining lifetime celibacy.[21]

It hardly helps that a sizable fraction of priests (perhaps as many as 40 percent) have a homosexual orientation. The reason for this high number is that many heterosexual men have rejected the clergy as a career because of its celibacy requirement. Homosexuals are less inhibited by this requirement and continue to be attracted by the notion of priestly fellowship. The struggle to maintain celibacy is just as difficult for homosexuals as heterosexuals. Even so, there have been many scandals about active homosexuals in the church, and some insiders allege that homosexual practices were widespread in their seminaries.[22] Due to the sex segregation of Catholic schooling, priests are likely to have more interactions with male students rather than females, so that boys are more likely than girls to be victims of priestly abuse.

Attempting to understand the motivations of abusive priests can help us to understand patterns of abusive behavior, but it hardly minimizes their breach of trust. The relationship between priests and their flocks is ideally one of reciprocal altruism. Such altruism is fragile. In the normal course of events, if one partner fails to live up to their end of the bargain they are ostracized. The vampire bat who refuses to feed her famished roost mate is kicked out of her perch and is liable to starve to death in a few weeks. Perhaps the most interesting aspect of the pedophile priest scandal is that those notorious individuals who have so conspicuously failed on their side of the bargain by undermining the spiritual welfare of their charges instead of promoting it have been allowed to continue with their priestly duties.

Most commentators now recognize that the Catholic Church itself faces a major crisis of confidence among parishioners because of its failure to punish sexual abuses. People who cover up crimes become accomplices after the fact. In this way, a few bad apples infect the whole barrel and the breach of trust of serial child molesters gets magnified into a general breach of trust between the Catholic Church and all of its members.

It is interesting that the worst scandals have emerged from the Boston Archdiocese, where support for the church, and for priests, had verged on hero worship. Many of Boston's Catholics are descended from poor Irish immigrants who relied on priests to help them get jobs, find housing, and educate their children, in addition to providing social services and spiritual guidance. Fanatical devotion to priests has meant that they received an unofficial status, similar to diplomatic immunity, so that their parking offenses and other minor violations were excused. For similar reasons, the sexual failings of priests were unlikely to be reported because parishioners admired their spiritual leaders, were oblivious and in denial of their crimes, and wanted to avoid harming them. All this has changed with the sensational publicity given to some shocking cases of serial child rapists whose criminal careers continued for decades with the tacit support of the church hierarchy.

How could the church have gotten so far off track? According to Vern Bullough, the Catholic Church has been willing to buy silence from victims of sexual molestation up to now because the number of priests involved is rather small.[23] By attacking child molesters directly, they are afraid of uncovering a much larger scandal of the many homosexual priests who are sexually active. There has thus been a balance of terror between pedophile priests and the hierarchy; exposing the former would bring ruin to the latter. This helps to explain why the church has been so slow to hand off its criminal priests to the civil authorities, who would like to lock them up out of harm's way.

This game of chicken came to a head in respect to the Reverend

Paul Shanley, a high-profile Boston priest accused of repeated acts of pedophilia.[24] His superior, Cardinal Humberto Madeiros of Boston, attempted to remove Shanley from his ministry. Shanley denied the charges against him and countered that the church's biggest problem was homosexuality, rather than pedophilia. He threatened to go to the press with claims about the prevalence of homosexual priests in the Boston Archdiocese and is reported to have said that if his allegations were fully aired, the Cardinal would have to dismiss many of his top clergy.

It is difficult to know whether Bullough's interpretation of these events is correct, much less to determine whether they apply more widely throughout the Catholic Church, but it is certainly plausible. Moreover, it helps to explain the curiously self-destructive lack of action by the Catholic hierarchy to shore up the social contract between church and laity that was so egregiously breached by a handful of pedophiles who have been allowed to run amok for decades.

To some extent, the church has been a victim of its own political correctness. There was an unwillingness to confront the prevalence of homosexuality in the church because to do so would risk being seen as homophobic. This is not just politically undesirable in an age of pluralism, it would also threaten the supply of new priests and compromise the vitality of American Catholicism in the future. Homophobia is offensive to the many priests who are themselves of a homosexual orientation. Over the past forty years, more than eighty priests in the Boston Archdiocese alone have come to the attention of church authorities as being in sexually abusive relationships, mainly with adolescent males, according to Cardinal Law, who released their names, belatedly, in 2002.[25] Their activities have rarely been brought to the attention of civil authorities even when they have broken laws by seducing minors.

It is easy to make abstract claims about tolerance of homosexual priests in the American Catholic Church, but the evidence for this is startlingly clear for anyone who wants to look. The most compelling case history is that of defrocked Boston priest John Geoghan, who has been described as the most notorious child molester in the history of American Catholicism. Geoghan was sentenced to nine to ten years for fondling a boy and was subsequently murdered in prison. He was the subject of approximately ninety civil sex-abuse lawsuits. Fifty of these suits were settled by the Boston Archdiocese for damages of over $10 million. Such huge costs of buying off victims of child abuse alienated major donors to the diocese's charities, many of whom began withholding funds in 2002. The intense adverse publicity surrounding the Geoghan case has also meant that the policy of trying to cover up priestly homosexuality and sexual abuse of minors will no longer work. Cardinal Law, who presided over the church's inadequate response to priestly sexual abuses, resigned under intense public pressure.

Although not as prolific as Geoghan, Paul Shanley, who is accused by over forty alleged victims, is more fascinating because of his very public endorsement of pedophilia.[26] Shanley's career as a public advocate of sex between men and boys is almost unbelievable for a man of the cloth who is ostensibly dedicated to chastity and the upholding of a biblical code of sexual morality that is hostile to gay sex of all kinds, much less that which constitutes a serious criminal act.

Shanley's life as a priest seems much more like the career of an antipriest because so many of his actions and ideas are in direct conflict with what we expect priests to do. He began, in the 1970s, with a very trendy ministry to Boston's street children that received much favorable publicity at the time. He specialized in sexually troubled youth. Once he had gained their trust, he allegedly molested them. According to diocesan files released under court order, the church has been aware of his sexual exploitation of children entrusted to his ministry since 1967. Shanley was accused of repeatedly molesting two six-year-old boys from his parish in 1983. Instead of defrocking him at the first signs of trouble, or at least alerting the police, his superiors ignored the abuses.

Shanley was active in the North American Man-Boy Love Association, an organization that devotes itself to the sexual abuse of children by adult homosexuals. He was present at its founding conference in Boston in 1987 and spoke publicly of the potential benefits to children of sexual relationships with adults. By 1990, Shanley had become something of an embarrassment to the Boston church and he was moved to California with a glowing recommendation letter from Robert Banks, a deputy of Cardinal Law. In California, Shanley and another priest opened a hotel for homosexual guests in Palm Springs that reportedly permitted gay sex beside the pool.

FALSE DOCTORS

In addition to the priesthood, medicine is a very highly respected profession with a reputation for dedication and hard work. In recent times, respect for doctors and other medical professionals has declined due to revelations that medical error is a leading cause of preventable death, possibly sending twice as many Americans to early graves each year as were killed in the entire Vietnam War. Many of the deaths are from hospital-acquired infections that could be prevented by proper hygiene. The Institute of Medicine (IOM) estimates that preventable adverse patient events, including infections, kill 44,000 to 98,000 people each year. In fairness to doctors, their numbers are guesstimates (based on the impact of improving hospital hygiene), and the IOM did not actually observe *any* hospital deaths.[27]

In addition to the problem of hospital-induced infections, we are just beginning to realize the harmful effects of many medications, particularly if used for long periods or in combination with other drugs. This problem is particularly acute for elderly patients, many of whom are sickened by chronic administration of several drugs whose complex interactions have not been scientifically tested.

Sloppy medicine is particularly distressing because most medical errors are preventable in principle. The problem is that the medical field has acquired such an aura of authority that doctors and other medical professionals cannot acknowledge their mistakes, which makes it rather difficult to devise systems that prevent errors from happening. Thus, many patients have been killed by receiving the wrong medication due to the illegibility of handwritten drug orders. Almost all such errors can be prevented by entering the information electronically and maintaining a computerized record of all drugs administered.

Medical errors are too frequent and it is reasonable to suggest that physicians bear some responsibility for their failure to improve the safety of their procedures. We do not tolerate aircraft mechanics who send planes into the skies with loose bolts, so it hardly makes any sense to tolerate physicians who preside over numerous damaging medical mistakes. That said, it is probably true that most errors occur despite good intentions. Doctors may not realize how often their written instructions are misread or misinterpreted by nurses and others, for example. They may continue to use faulty procedures because they are unaware of just how error-prone they are.

Doctors thus do a great deal of harm but they may be generally unaware of their errors and so believe that they live up to their Hippocratic obligation to do no harm to patients. Even so, there are some rare but well-documented cases of doctors and other medical professionals abusing the trust invested in them by deliberately harming their patients.[28] This is spectacularly true in the case of doctors recently identified as serial killers.

England's most prolific serial killer was Dr. Harold Shipman, who ran a busy practice in the sleepy town of Hyde for twenty-four years until his conviction of murder in 2000.[29] Recent analyses of death certificates he wrote suggest that he murdered approximately 250 people, mostly by injecting them with apomorphine. Dr. Michael Swango is one of the most prolific of American serial killers. The FBI believes that he may have murdered approximately sixty people.[30] Then there is Efren Saldivar, a respiratory technician, who may have killed over one hundred of his patients by asphyxiation and poisoning.[31] Each of these notorious individuals has recently been convicted of killing a small number of victims, and Shipman committed suicide in prison in 2004. An ex-nurse, Lynn Majors, was also convicted of killing four of his patients in Springfield, Massachusetts.[32] There are many historical examples of homicidal physicians as well, one of the most egregious being Josef Mengele, who not only was an active participant in the Holocaust but also killed scores of Jews in gruesome, racially motivated medical experiments.[33]

The fact that doctors can be so successful at killing their patients with impunity is shocking for exactly the same reason that some priests have continued for decades as serial child molesters. These crimes appall because they involve such a deep betrayal of trust. The amount of trust invested in priests is one reason that they were able to abuse children with impunity. Trust has also been a major ally of serial killing by medical professionals. In Dr. Shipman's case, the esteem with which he was held in the community of Hyde prevented local people from noticing that large numbers of his patients died suddenly without a previous illness that could adequately explain their demise.

It is difficult to say if there are more serial killers in medicine than in other professions. In all probability, such creepy individuals are as thin on the ground in the medical profession as they are elsewhere. The few homicidal medical professionals receive a great deal of sensational publicity because their social position of trust sometimes allows them to escape detection or prosecution. The upshot is a large body count that makes for sensational reading.

When criminal attorneys try a homicide case, they often dissect their arguments into questions of means, motive, and opportunity. Doctors have constant access to means of killing others, specifically lethal poisons, that may be difficult to detect, even in an autopsy. They not only have access to such means of destruction but understand how to use them effectively and discreetly. What is more, they see patients alone, which gives them many opportunities to carry out their criminal intentions.

This brings us to the third issue of motive. Many people find it hard to understand why someone who has dedicated his life to medicine would consider violating his Hippocratic oath and killing the very person he is hired to protect. Analysis of medical serial killers suggests that there are two major motives, mercy killing and thrill killing. In some cases it may be difficult to tease these motives apart.

Mercy killing has long been looked upon with horror and suspicion in the medical community and by the public. Yet, there are at least two places in which killing of patients by physicians has been legalized. The state of Oregon has legalized physician-assisted suicide in which doctors supply patients with a lethal poison and help them to administer it. Doctors in Holland are also allowed to terminate the lives of their patients.[34]

Very few Oregonians have chosen to commit suicide with their doctor's help. Most of them had suffered from painful incurable illnesses, and the ending of their lives was clearly voluntary. Euthanasia is legal in Holland, and it does not require voluntary consent. The majority of Holland's elderly population fears that if they must go to a hospital, they can be put down involuntarily. Dutch law requires the keeping of careful records on euthanasia but busy doctors rarely provide the necessary documentation. One reason that Dutch law gives physicians so much power over the life and death of their patients is that people often build up a lifelong trusting relationship with their family doctor and thus are inclined to trust all medical professionals very highly. Another issue is the rather communitarian views of most Dutch people, specifically the view that valuable medical resources should not be wasted in heroic attempts to keep people alive who will expire in a few days anyway. Better let the elderly die and invest medical resources in younger patients who can benefit a lot more in terms of years saved and quality of life. Doctors thus have the responsibility of deciding when life support systems should be shut off. They generally consult with next of kin but sometimes make the decision on their own. For patients who are not on any life support system but who are in great pain from a terminal illness, or actually approaching death, life may be shortened by the administration of high doses of opiate drugs. This was also the method chosen by Dr. Shipman to dispose of his healthy patients in Hyde.

Callous disregard for the interests of elderly people is one possible consequence of Dutch legalized euthanasia. Involuntary killing of elderly patients excites a great deal of moral outrage in other countries, and it is not a feature of life in most indigenous societies studied by anthropologists. Thus, the !Kung care for their elderly sick, providing them with nourishing food, healing rituals, entertainment, and social support.[35] Many suffer from eye diseases so that they are unable to help with children or food preparation, much less hunting or foraging. Not all societies are so kind to their elderly, however. Thus, during times of food scarcity, the Eskimos have a tradition according to which the aged are expected to depart on ice floes, thereby sealing their own doom while altruistically ensuring that there is more food available for younger adults and children.[36]

The Dutch system of euthanasia seems to have operated reasonably well, but dead people tell no tales. The system is founded on complete trust. Elderly Dutch patients worry about whether hospital doctors will cure them or kill them because these specialists are relative strangers. A system that is based on implicit trust can work well provided all the relevant physicians live up to that trust. One appalling possibility that the Dutch have evidently not considered is that their system is wide open for abuse by a serial-killing doctor. At present, they have no way of distinguishing between a malicious serial killer and a conscientious doctor. Even if widespread violations of the law occur, they have no records with which to prove that crimes were committed. Murphy's Law predicts that a very prolific serial killer is, or will be, active in a Dutch hospital specializing in geriatric medicine.

This horrifying prediction is not so improbable when you consider that one of the medical serial killers in the United States specialized in disposing of elderly patients who were terminally ill. Efren Saldivar told police that he helped to kill at least one hundred people because he could not bear to witness their suffering.[37] Initially, at the age of nineteen years, he allegedly began suffocating patients by connecting ventilator tubes to cut off the flow of oxygen. Then, inspired by a television story about an "angel of death," he reportedly took a more proactive role and began poisoning his patients using morphine (an opiate) or Pavulon, a paralyzing drug. It is interesting that opiates are often prescribed to reduce the suffering of people who are dying and have the unintended consequence of hastening death. Normal palliative care thus has an element of euthanasia, even in countries where euthanasia, as such, is illegal.

Although Saldivar has been convicted for some of his crimes, his motives remain questionable. According to his own account, he was moved by pity for dying patients and claimed that he carefully selected his victims, preferring those who were unconscious and had "do not resuscitate" orders on their medical charts. He also said that he preferred to kill only those who looked ready to die. Saldivar's actions hardly differ from those of hospital physicians in Holland.

He was not a physician, however, but a respiratory therapist, and he was not in Holland. Mercy killing could well be motivated by the thrill of killing other people.

The most heinous modern serial-killing physicians, Shipman and Swango, were clearly motivated by the thrill of killing. In Shipman's case, there was never any admission of guilt, so that there is no record of his own account of possible motives. The very lack of a credible motive may be one reason that he escaped detection for so long. When Shipman yielded to the temptation of greed, he made mistakes and revealed himself. Thus he forged the will of one of his victims, eighty-one-year-old Kathleen Grundy, leaving the entire $600,000 estate to himself and cutting out her relatives. The signatures of witnesses were also fabricated, which proved to be Shipman's undoing because the witnesses denied seeing the altered will.

Although Shipman specialized in the killing of elderly widows, he did so because they were targets of opportunity: more elderly women live alone because women generally outlive their husbands. He was no mercy killer: some of his victims were middle-aged, and most were in reasonably good health.

Shipman's success as a serial killer was based on the trust and personal respect that he enjoyed in the local community.[38] He also exploited a wrinkle in the law which stipulated that an autopsy was unnecessary if a person had recently been seen by her physician. This meant that the opiate drugs used to kill patients went undetected.

As far as motivation is concerned, Michael Swango[39] is the clearest example of a serial killer who killed for the sheer pleasure of it. In many ways, he is a classic serial killer whose odd obsessions and behavior brought him adverse attention from colleagues from the beginning of his medical career. Shipman was much more careful and discreet and had never been in trouble, apart from a single charge of illegal possession of morphine for personal use.

Swango's behavior was peculiar from the very beginning of his career, in the early 1980s. It was odd enough to attract the attention of his residency review committee at Ohio State University, in Columbus, who did not extend his employment there. A string of suspicious deaths occurred and Swango was permitted to leave quietly despite a murder investigation. Evidently, university administrators preferred to distance themselves from the adverse publicity and they reportedly blocked the investigation against Swango. From this inauspicious start, Swango made numerous career moves, as colleagues saw through his superficial charm to the pitiless monster within. Working hard to outrun his past, he obtained employment in various hospitals in South Dakota and New York, as well as Zimbabwe, Zambia, and Saudi Arabia.

Swango obtained each of these positions despite being investigated for murder and actually being convicted of aggravated assault in the deliberate poisoning of paramedic coworkers, a crime for which he did jail time in Illinois. He obtained jobs through a creative use of false documents, aliases, and phony employment histories. On some occasions, he acknowledged the criminal conviction but lied about the nature of the charges. He was arrested in June 1998 while attempting to reenter the United States and was convicted on fraud charges connected to his employment in New York, for which he received a forty-two-month sentence.[40]

There can be little doubt about Swango's motivation in harming his patients. He was not only a classic serial killer but openly identified with Ted Bundy, a serial killer of girls and young women, whom he raped and mutilated. Swango liked to read novels about doctors who were murderers. A passage from one of these novels, *The Torture Doctor*, was hand copied by Swango into a notebook. It provided powerful ammunition for the prosecutors when he was charged with killing a nineteen-year-old gymnast, Cynthia McGee, who was hospitalized at Ohio State University Hospital following a traffic accident. It read: "He could look at himself in the mirror and tell himself that he was one of the most powerful and dangerous men in the world . . . he could feel he was a god in disguise." Swango was a textbook example of a thrill killer. Like Shipman, he killed his patients purely because doing so gave him a sense of the power of life and death over others.[41]

Organized social groups of every kind rely heavily on the willingness of individuals to follow rules of expected behavior and generally have some kind of enforcement system to ensure that they do. Public conformity is a natural tendency, but, according to social psychologists, when given an opportunity, individuals are liable to pursue their own interests to the detriment of the group, as spectacularly illustrated in serial-killing medical professionals. This is not just true for murderous physicians and sexually abusive priests who feel themselves to be above the law, it also applies to the arrogant greed of crooked company officers, for example. Witness the recent feeding frenzy among corporate executives, in companies like Enron, where the greed of the officers ran the company over the cliff of bankruptcy.

Business ethics is interestingly complex because the whole philosophy of business enterprises is based on the self-interest of owners. Corporate officers are expected to rein in their own greed, because they are paid well for their services. In America, executives have received ever-higher compensation packages, in a frenzy of greed that rivaled the California gold rush of 1849. In the end, the huge salaries, the stock options, the interest-free loans, and the company credit cards served to incite greed rather than to satisfy it. Whatever else is said about Enron-type frauds, they are among the most conspicuous cases in history of company officers feathering their nests at the expense of their employers, the company stockholders. There are many other cases of individual selfishness undermining the mutual benefits of altruistic groups. One is the phenomenon of road rage.

ROAD RAGE: WHEN ALTRUISM FAILS IN DRIVERS

Social animals can benefit from their combined effects on the environment, particularly in respect to their joint architectural efforts. Thus, bees build honeycombs whose cells are used to store food surpluses for lean times as well as to house succeeding generations. For many social insects, like termites, the most impressive construction project is the colony in which they live. The termite hill is a protected environment that is strong enough to resist the intrusion of many predators. It is also a controlled environment where temperature and humidity fluctuate only in a narrow range that promotes growth and survival of colony members.

Human heating, air conditioning, and ventilation systems are designed to fulfill the same basic functions. Despite their complexity, they are generally not as effective as the climate control system inside insect colonies. The temperature inside a termite hill rarely changes more than about two degrees Centigrade in the course of a day, whereas the temperature inside a room at a particular spot often fluctuates more than five degrees. Thus, it gets much cooler over the vent when the air conditioner is running than when it is not.

Much of the environmental constancy inside an insect colony is due to good design.[42] Animal architecture has its limitations, however. During the summer, the colony heats up and bees have to turn on the fans—themselves. Maintaining favorable environmental conditions for the insect society requires behavioral conformity, and many bees fan their wings together to circulate cooling air through their residence.

Just as bees coordinate their activities to defend their microclimate against dangerous fluctuations, so they must regulate their movements so that individuals are not forever colliding and working at cross-purposes. How this is done is somewhat of a mystery. When you look inside a social insect colony, you see what looks like chaos but it is organized chaos nevertheless, analogous to the sound of an orchestra tuning up before the performance. Each individual is doing its own thing, but in a way that is coordinated with the actions of others. Foragers leave slowly and steadily, rarely colliding at the narrow entrance of the hive.

The regular movement of traffic on a human highway provides an obvious analogy. There are two basic ways of accounting for such order. One is that it is an inherent property of the physics of moving objects in a restricted space. Engineers have thus developed computer models that are rather good at mimicking the flow of traffic through cities. The same kind of logic has been applied to activity patterns in social insects, specifically ants of the genus *Leptothorax*.[43] When there are only a few ants, they move around more or less randomly, erratically taking care of the young. When the colonies become larger, they get more organized. Now the ants adopt a regular rest-activity cycle of about twenty-five minutes, almost as though they had joined a union with mandatory coffee breaks! It was long assumed that this clocklike regularity had to be due to an internal clock, but computer models have shown that simulated ants following a few simple rules of interaction develop a similar rhythm. The order is thus intrinsic to the physical system itself. The relevance of these conclusions for human traffic regulation is that urban traffic will generally be a great deal more highly regulated than the movement of traffic on a lonely country road. Another way of saying this is that our driving must be a great deal more sensitive to the movements of others in cities than it is in the countryside.

This is a bottom-up approach to traffic regulation. The complementary top-down approach focuses on the rules of the road, already discussed as an example of altruistic conformity. The presumption is that people learn the rules of the road by a combination of study and experience and then obey them to the letter, promoting an orderly flow of traffic. Everyone knows that in reality minor infringements are the order of the day. The roads are crammed with amber gamblers. On some highways, lane-change signals are frequently omitted. Frequent lane-changers would be dead if they lost time switching their turn signals! Drivers who remain below the speed limit will also get honked off the road in many cases as an irritating impediment to the free flow of traffic.

Many drivers thus routinely violate the rules in a way that increases danger for themselves and other motorists. These are calculated risks, however, and they can increase the flow of traffic. Thus, many city drivers are guilty of tailgating, driving too close to the driver in front for optimal safety, but doing so helps traffic to flow more rapidly. In fact, if city drivers did not violate the rules in this way, metropolitan areas would be forced to build new roads to accommodate the backed-up traffic. In this case, it seems that the physical properties of the traffic system win out over the rules of the road. Impatient drivers may pose an increased risk for collisions on the roads, but they also contribute to the efficiency of the system. There is thus some ambiguity as to whether they harm, or help, the common good of the highway system. The conclusion would clearly depend on a precise measurement of increased danger versus increased efficiency and some method of equating the benefits and costs to the community.

No such ambiguity surrounds the phenomenon of road rage, in which angry drivers pose a deliberate threat to other motorists.[44] According to some experts, road rage may be a factor in as many as two-thirds of deaths in road accidents, although such estimates are highly debatable. If obedience to rules of the road is altruistic, there could be no clearer demonstration of the failure of human altruism than road rage, in which drivers become so angry during their interactions with other motorists that they behave in very dangerous and threatening ways, sometimes attempting to run a complete stranger off the road, or attacking them with homicidal intent.[45] If people have an evolved predisposition toward altruistic behavior in groups, why does road rage occur?

One way of explaining this apparent contradiction is by recognizing that conformity to complex rules can be difficult and stressful, thereby provoking emotional responses that contribute to aggression. Researchers have shown that people who succumb to road rage are often not very good drivers. Drivers who admitted to angry and threatening behavior on the road, such as deliberately cutting others off, tailgating, and chasing other cars, were more likely to have had a serious car crash in the past and to have had a minor collision in the previous year. They were also more likely to drive while intoxicated and to regularly exceed speed limits.

Hostile drivers also suffer more from psychological disorders than other drivers. Many have antisocial personality disorder, a comparatively rare problem that makes it difficult to conform to social rules and obey laws. The underlying problem is emotional—a lack of empathy with other people and a comparative inability to feel guilt over one's misdeeds. Antisocial personalities are biologically predisposed to be defectors, or cheats, in social contracts. They used to be referred to as moral imbeciles on account of their limited understanding of how social contracts work. Antisocial parents are liable to abuse and abandon their own children, suggesting this is a true disorder, in the sense of undermining reproductive success. People who succumb to road rage are more likely to suffer from other psychological problems as well, including an inability to control outbursts of rage.

Road rage is three times more common for people under thirty years than it is for people over fifty-five. It is also more common for males than females. The underlying reason is that young people, and men, are more impulsive and reckless than older people and women. For men, at least, the age difference is likely due to declining levels of testosterone with age.

People with anger management problems and other personality disorders, not to mention impairment by alcohol, are more likely to fall apart under the psychological stresses of driving in difficult conditions. They seethe with anger and often misinterpret driving errors by others as deliberate acts of hostility directed at them.

The view that the difficulty of conforming to traffic regulations is increased for those individuals with psychological and cognitive impairments clearly does not apply to the elderly, who may have greater problems handling ambiguous traffic situations but are generally calmer and slower to anger. If road rage is really a problem of information overload undermining altruistic conformity, then it should be more likely to occur in ordinary people when they are especially hurried, or when traffic conditions are unusually difficult.

There is some support for each of these predictions. A recent survey of South African truck drivers found that those with very tight delivery schedules were more likely to have bad driving habits and to drive aggressively. If drivers are perpetually in a hurry, then even minor delays can build up their level of frustration so that they drive more aggressively.[46] Studies using driving simulators find that normal people become angry in frustrating driving situations such as being stuck behind a truck on a narrow winding road, or driving in a three-lane urban highway that has slow-moving traffic. Comparisons of ten US cities found that Boston had the highest level of road rage. This may have nothing to do with the disposition of Bostonians and everything to do with the congested and chaotic nature of traffic bottlenecked in the seething heart of this metropolis.[47]

When people are delayed uncontrollably by traffic jams, they become frustrated and angry so that they are more likely to vent their rage at the perceived mistakes of other motorists. Impatience, or time-urgency, undermines altruism. One of the most celebrated, and humorous, demonstrations of this involved a seminary study of the willingness of trainee priests to help strangers in distress.[48] As they walked from one campus building to another to deliver a sermon, they encountered a man stretched out on the path as though he had just collapsed. When the seminarians had plenty of time to kill, most stopped to minister to the stranger. When they were late for their sermon, most walked right by him, in some cases even stepping right over him. The savagely humorous part of the research was the topic of their sermon: the Good Samaritan.

Altruism toward other individuals can thus break down under fairly predictable conditions involving psychological stress and conflicting demands on a person's time and attention. Economists have identified a more generalized failure of altruism, namely, the propensity of individuals to degrade public property through selfish abuses. The problem of individual selfishness undermining the public good is often discussed in terms of a metaphor, or thought experiment, known as the tragedy of the commons.

THE TRAGEDY OF THE COMMONS

The tragedy of the commons, or land held in common, is overgrazing. Biologist Garret Hardin generalized this problem to all manner of social situations in which there is a conflict between the needs of the individual and the welfare of the group.[49] A person who throws a candy wrapper onto a previously unspoiled wilderness detracts immensely from the amenity value of that place for other hikers. Similarly, industrial polluters degrade the more critical shared resource of the air we breathe with adverse health consequences.

Hardin, writing in 1968, was struck by the fact that some of humanity's most dogged problems involve scenarios in which the selfish behavior of the individual provides him with a small immediate benefit but contributes to the irreversible destruction of a shared resource. Thus, the wilderness hiker who discards a small sticky candy wrapper on a hot day may be appalled to discover glass bottles, beer cans, and used diapers on his next visit and may decide never to go there again.

Land was held in common by medieval English villages; ranchers in the American West also grazed publicly owned range lands. Such commons systems may work well for extended periods of time, so long as the population size is kept below the carrying capacity of the environment (the maximum number of grazing animals that the pasture can support). In the English case, population growth remained low due to the impact of diseases and warfare so that each herdsman continued to graze his allotted number of animals without damage to the grassland. When the number of grazing animals reaches the carrying capacity of the pasture, the tragedy of the commons manifests itself. Logically speaking, it would be in the interest of each herdsman to graze fewer animals, thereby preserving the fragile grassland ecology on which they all rely for their livelihoods. The problem is that the benefit to the commons of each herdsman altruistically refraining from putting an animal out to graze is distributed throughout the community, but the cost is borne by him alone. It is another case of the prisoner's dilemma in which altruistic individuals are liable to be taken advantage of by others. According to Hardin, the herdsman can be expected to play his role in the tragedy, putting out as many grazing animals as he can get away with. The inevitable result is overgrazing, followed by soil erosion, colonization by weeds, and the irreversible destruction of the commons.

Hardin considered that the degradation of public property by selfish individuals was more or less inevitable, unless arrested by some kind of governmental regulation. One solution is privatization. The English common lands were enclosed for private use, for example. Another is the implementation of government regulations that prevent or penalize acts that damage public property. Thus, the problem of overcrowding in inner cities has been lessened by bans on parking in busy areas and the use of parking meters to discourage motorists from leaving their cars for unnecessarily long periods. The problem of environmental pollution is addressed by a detailed system of fines and penalties, as well as the implementation of many laws governing emissions from private vehicles, as well as industrial smokestacks.

The argument that when people are left to themselves, their base natures will wreak havoc, unless controlled by government regulation, is an ancient one. In religious terms it reverts to the argument of original sin. As noted, English political philosopher Thomas Hobbes developed a political version of the same idea in the seventeenth century, arguing that government is necessary to save us from the inevitable conflict produced as we try to satisfy our base appetites.

Garret Hardin argues that one inevitable consequence of economic development is the loss of commons systems. One of the earliest examples of abandoning the commons system is food production. Instead of gathering food from the forest, modern humans grow their food crops in enclosed farms. Hunting and fishing are no longer a guaranteed and unfettered right: instead, they are subject to precise governmental regulations that stipulate when a particular species may be taken, in what quantity, and by what means. The commons is no longer considered an appropriate dumping ground for domestic sewage and industrial waste. Governments regulate the spraying of toxins into the environment in the form of weedkillers and insecticides. Hardin noted that there was very little regulation of noise produced in public spaces, and this is still largely true. One exception is the mandatory use of mufflers on cars but not on other bothersome noises from radios, lawnmowers, jet skis, and even aircraft.

Hardin is a strong proponent of government regulation because he believes that it is the only way to save ourselves from the destruction of common resources as a consequence of their selfish use. If you are offended by billboards that give your local highways a trashy feel, then you must endorse government regulations that ban such unsightly advertisements, as has been done in the state of Maine. A similar logic would ban advertising on commercial radio and TV, but this has not been done, presumably because such enterprises rely on advertising revenue for their very survival.

Hardin's view, that more government regulation is the only answer to commons problems, has not gone unchallenged. It is certainly true that more government is a predictable response to the environmental problems that follow rapid economic development and population increases. It does not follow, of course, that governmental regulation is the only solution to commons problems. Our hunter-gatherer ancestors had little to fear from overpopulation because their numbers grew so slowly. Nevertheless, they could theoretically have experienced problems from overhunting of game animals by neighbors. This rarely affected them, however, because they enjoyed exclusive hunting rights over a large range that was adequate to their needs. According to anthropological research on existing societies, such hunting rights are enshrined in custom and do not require any kind of government enforcement. Science writer Matt Ridley also points out that the regulation of lobster fisheries in some parts of Maine is by means of a hereditary system of lobstering territories.[50] This system is fascinating because it is not supported by any property laws. As far as the law is concerned, anyone can go into business as a lobster fisherman. In reality, those whose unofficial, hereditary rights are infringed on will likely retaliate, either by destroying the interloper's equipment, or by threatening his boat or his person.[51]

Such bottom-up methods of property regulation are by no means peculiar to Mainers, or even to human beings. Thus, farmers in Kenya have no legal title to the land they work and maintain their ownership of the land by keeping family members in residence. The larger their families, the more land they can work and maintain. It is hardly surprising that agricultural Kenyans have large families and equate children with wealth.

Such preoccupation with territories that can be used to raise families is far from being a human trait, of course. Many male birds defend territories during the breeding season. When females select a mate, they often seem more concerned with the quality of the territory than they are with the quality of the mate. Animals as different as ants, flies, fish, lizards, wolves, and monkeys all defend territories that contain food sources and/or shelter and refuge from predators. In each case, territoriality promotes spacing and prevents overpopulation from destroying the habitat of a species. Government is not always necessary to solve problems of competition over valuable resources.

In simpler societies, at least, the commons can be successfully partitioned through nongovernmental means. Informal adherence to group territories among hunter-gatherers are an efficient solution to potential conflicts of interest and to the potential for overhunting. But what about the more complex societies in which we live today?

PART 4

Kindness and Politics

CHAPTER 10

Tapping Human Altruism

Students of human altruism are confronted by the paradox of very unpleasant actions, like warfare, being derived from altruistic motives. Altruism within groups makes people both happier and healthier but has devastating consequences when it fuels aggression against out-groups. Can we anticipate a future in which intergroup hostilities are less probable? Can we live down our history of endless wars and seemingly futile bloodshed? Here we grapple with the problem of psychological adaptations for cooperation that can either unite strangers or stir up international conflicts. One obvious solution to the problem of global conflict is a world state, but a single government for the entire world is a frightening prospect, vulnerable as it is to despotism and totalitarianism. A more acceptable solution would be to live in a world where independent states recognized common global problems, like international terrorism, and cooperated in solving them.

THE ORIGINS OF WARFARE

Lifestyles of extant hunter-gatherer societies approximate the Garden of Eden myth, that is, a society with little organized conflict or warfare. This evokes the noble savage idea of French philosopher Emile

Rousseau (1712–1778), but the noble savage is a myth. Hunter-gatherers are not particularly admirable. In fact, they are far more violent than urban people because there is no formal legal system that can be used to resolve disputes, apart from arbitration by the headman. Physical aggression is thus a primary means by which individuals protect their interests. We think of societies like the !Kung of the Kalahari Desert as peaceful, even though they do occasionally make war with their neighbors. But homicidal violence is quite common there, much of it motivated by sexual jealousy. Men kill other men who have affairs with their wives. Sometimes they beat their wives severely to intimidate them from having future affairs.[1]

The real difference between hunter-gatherers and modern societies is that the former lack the political organization that would elevate their disputes beyond the level of squabbles between individuals or kin groups. However, with increased political complexity comes larger groups and the possibility of large-scale warfare.

Societies differ greatly in their predilection for making war. Such differences are illustrated by Native American tribes. The Shoshone were peaceful, never attacking their neighbors, but the Paiute were in a state of constant warfare with adjacent groups. Both groups inhabited the Great Basin region of the West and both subsisted mainly on wild grass seeds. The critical ecological difference was that the Paiute lived on the banks of rivers where grasses grew plentifully for long periods. The Shoshone inhabited drier short-grass prairie where there were no permanent streams. This meant that their food was scattered and lasted only for a brief time. The Paiute settled on stretches of river where food grew plentifully whereas the Shoshone were nomadic, moving constantly in search of food.[2]

Being settled, the Paiute naturally defended their food sources from neighbors on either side of their stretch of the river. If their numbers increased, they encroached on the territory of neighbors as a matter of necessity. This conflict over land was at the root of an endless cycle of attack and counterattack among competing groups. Similar patterns are seen around the world when population expansion intensifies competition over land.

The root of the Paiute warfare was not just conflict over food but the fact that their food sources were concentrated and permanent: well worth defending. The Shoshone searched for food in small family groups and they never fought other family groups they encountered while foraging. Food was located in small ephemeral clumps that survived drought only for short periods and were just not worth defending.

The fact that clumping of food sources makes them defensible and thus increases aggression is true not just of humans but for various other species as well. In fact, it was inadvertently demonstrated by researchers in primatology who used to leave out food to attract their subject animals, thus saving them days of thankless trekking through the wilderness in search of their target species. The food worked but created a problem: it made the animals highly aggressive, distorting their natural behavior so much that the purpose of the field research was undermined. In fighting over "free" food, primates often become aggressive enough to cause each other serious injury. It is now considered unethical to serve provisions to attract wild primate populations in the course of a field study.

The effects of providing food in relation to aggression in wild animals is also familiar to anyone who has ever kept a bird feeder. The feeder attracts a number of species that would not normally feed together, and there is a great deal of jockeying for position that would not otherwise occur in nature.

Heroic defense of large comparatively unrelated human groups in modern warfare can be explained in terms of reciprocal altruism, as discussed earlier. Reciprocation occurs across kin groups, rather than individuals. In other words, kin groups who are willing to sacrifice some individuals in defending their shared interests with allied kin groups are more likely to prevail over aggregations of kin groups that have no such sacrificial tendency. When vital interests, such as essential food sources, are at stake, an effective joint-kin group defense is called for, and this defense may be costly.

Among animal societies, something analogous to human warfare is seen in defense of a group territory, as seen among wolves, for example. The costs of such defense can be explained in terms of kin selection. By defending their territory, animals are effectively defending their offspring, or other close kin who need a secure place to live and find food.

Apart from humans, many other social mammals, such as lions, hyenas, monkeys, and chimpanzees, defend group territories. In some cases, the boundaries are scent-marked, and individuals straying into foreign territories do so at their own peril. It is quite common for a lone wolf, or a solitary lion, to be attacked and killed by the group whose territory is infringed upon. When groups meet at a shared boundary, hostilities are liable to break out and, in some cases, individuals may be killed.

From an ecological point of view, the comparison between group aggression among territorial mammals, like monkeys or hyenas, and group aggression among territorial human groups is fairly compelling. Needless to say, the underlying psychological processes can be very different. Humans employ a great deal of planning and strategy in these matters, rather than a stereotyped aggressive response to strangers at the border of a territory. Yet, this distinction no longer seems as rigid as it used to. In one notorious case, a band of chimpanzees systematically attacked and killed each of the males in another group—a campaign that is remarkably similar to human tribal warfare. Such apparent planning is not seen in the group aggression of most other species, of course.[3]

The existence of something like war in other species is interesting because it contradicts the common view that organized aggression is a peculiarly human form of depravity. Pessimists often assume that warfare gets worse at higher levels of economic development, also.

This conclusion may well be true for the first step up the ladder of economic development, from hunting and gathering to food cultivation. Anthropologists conclude that warfare is extremely common in horticultural societies, where much of the food is cultivated. Many of these societies inhabit tropical regions of South America, Borneo, and New Guinea. Most combine a diet of cultivated crops that are high in carbohydrates, including taro, plantains, and sweet potatoes, with protein from the meat of hunted animals.[4]

In contrast to hunting and gathering that permits only a very low density of population, tropical gardens are highly productive in the first few years of cultivation (after which the soil becomes seriously degraded and has to be abandoned). Instead of the band of about forty individuals characteristic of a hunter-gatherer group, tropical horticulture supports populations of several hundred people who subsist primarily on the food grown in small gardens adjacent to the village. Increasing village population depletes game animals from a much larger region, however. Scarcity of game animals increases the likelihood that neighboring villages will go to war. Among the Mundurucu of Brazil, head-hunting raids are launched against neighbors whenever the peccary, a wild pig that is a favorite game animal, becomes hard to find. The Mundurucu are themselves well aware of the connection between warfare and competition over game animals. A hunter returning with the severed head of an enemy is honored as a "mother of the peccary."[5]

How violent are horticultural societies? One case is the Yanomamo of South America, whose antics are well known on account of the documentary films and books of Napoleon Chagnon of the University of California at Santa Barbara. In his well-read book, *Yanomamo: The Fierce People*,[6] the subtitle not only implies that they are unusually aggressive but also depicts these people as they see themselves. It is estimated that approximately one-quarter of men die in battle. This astonishingly high mortality is actually quite typical among tropical horticulturalists. Other groups, like the Mundurucu of Brazil, are much more violent than the Yanomamo, with over half of the men dying in warfare.[7]

These numbers are surprising in view of the simple state of military technology among such indigenous peoples. Battles are often comparatively small-scale, consisting of no more than a surprise attack on a single individual as he leaves the village to relieve himself. What they lack in technology, and scale, they make up for in persistence.

People like the Yanomamo are intelligent and self-reflective, despite their penchant for violence and seemingly gratuitous cruelty to animals and other people, including appalling violence against women as reflected in frequent injuries to wives and brutal gang rapes of captive women. They are tragically aware of the futility of warfare, with its endless cycles of attack and revenge, recognizing it as a vicious cycle out of which they cannot escape.[8]

For the Yanomamo and other such groups, once hostilities begin, it is impossible to get the genie back in the bottle. When attacked, a group must mount a spirited defense, or be forced to find refuge in less desirable habitat. Instead of sitting around waiting to be attacked, it is better to carry out a preemptive strike that weakens and intimidates potential attackers.

The tragic futility of the cycle of violence has impressed all observers of prolonged ethnic warfare, including that in industrial societies. Difficult as it may be to resolve conflicts like those in Northern Ireland, Bosnia, and Israel, it is always possible to appeal to a higher governmental authority to broker a peace agreement. For indigenous societies of the past, there was no higher authority than the village headman, and thus no means of arbitrating, and resolving, conflicts. The conflict problems of societies like the Yanomamo can be simply stated. They lack a higher authority to mediate disputes and, for various reasons, find it difficult to make a lasting peace themselves, even though everyone would be much better off in the long run.

The Yanomamo are caught on the horns of a prisoner's dilemma. Their previous history has been full of defections, in the form of treacherous opportunistic attacks that often involved temporary coalitions with other villages. Such treacherous attacks are not forgotten but kindle a burning desire for revenge. In a world where defection is common, and expected, it is naive to imagine that peace pacts can solve the problem of warfare. Villages are stuck in a Hobbesian war of every village against all others. They do not like this situation, seeing it as futile and tragic, but they are trapped by the dilemma that altruism cannot exist in a world where the selfish interests of the actors are in conflict, and where selfishness is rewarded, at least in the short term.

All cycles of war and revenge are tragically futile, but the futility of war has an added dimension in the Yanomamo case. Yanomamo communities suffer from a scarcity of women, and the explicit purpose of many of their raids is not to drive competitors out of their hunting territories but to capture women who will serve as their wives. Women are noncombatants in Yanomamo warfare, and defense of a village thus requires the highest possible number of males. Girls are less valued than males and infanticide of females is common, which means that there is a scarcity of females of reproductive age in each community. Hence the motivation for launching raids to abduct young women. Warfare thus reduces the number of women, and the scarcity of women increases the propensity for making war. The Yanomamo are trapped in a futile cycle of killing baby girls and then going to war in an attempt to replace their lost females.[9]

Horticultural societies are clearly more warlike than hunter-gatherers. Whether their conflicts are over land, game animals, or women, they have much to fight about. Armed conflicts are entered into to solve the problems of the belligerent group, but they inevitably lead to a great deal of pointless, yet unavoidable, suffering on both sides.

In very general terms, the increased bellicosity of horticulturalists is due to increased population density and the increased food production that makes this possible. Does increasing social complexity inevitably boost warfare?

DOES ECONOMIC DEVELOPMENT INCREASE WARFARE?

Most social creatures can accommodate themselves to groups of varying sizes. This is obviously true in the case of social insects where large colonies begin as a single pair of reproductives. Students of ant behavior find that individuals accommodate their activities in interesting ways to the increasing numbers of ants in the colony, adopting a regular cycle of activity and rest when the group increases to a certain size, for example.

There is much that remains unknown about adaptations for group living in ants, and the same is true for our own species. Nevertheless, with increasing social complexity, defined in terms of the number of individuals encountered each day, there is a necessary increase in behavioral conformity if the society is to function efficiently. From this point of view, behavioral conformity is a kind of altruism that benefits the entire community, even where there are immediate costs to the individual. If a person stops at all of the red lights, for example, she may get home from work later, but the traffic system is much safer as a result, and everyone benefits.

With increased economic development, and increased social complexity, greater conformity is required. Does this make us more vulnerable to dying in war, or does it protect us from violent death? Has state-level organization increased our likelihood of dying in war compared to that of tribal societies? There is no avoiding the fact that modern military technology facilitates the destruction of large numbers of people in a very short period of time. The body counts of the war to end all wars, World War I, and the much deadlier World War II have already been mentioned. World War II was also associated with the Holocaust—a horrifying product of Nazi conformity—and with the first use of nuclear weapons. Given such historically recent developments, it might appear that we live in a highly dangerous world in which the very triumphs of technology and social organization that permit the affluent lifestyle of modern countries are also a grave threat to each of us.

The spectacular scale of the horrors of World War II, not to mention the atrocities perpetrated by al-Qaeda on American civilians, makes us inclined to see modern times as especially war-torn, bloody, and depraved. During the twentieth century, over 100 million people died from war-related causes (including disease and famine). This is a huge figure. Averaged out over the population of the planet, it becomes less frightening. Despite the horrifying scale of its wars, only about 1 percent of people died due to warfare in the twentieth century.[10] This is nothing to brag about until one makes a comparison with preindustrial societies. According to Lawrence Keeley in *War Before Civilization,*[11] one person per two hundred died in war *each year* in tribal societies. This is *twenty* times the mortality rate of developed countries. If we had an equivalent war mortality, two *billion* people would have died in war during the twentieth century instead of 100 million. Even though complex political entities like nation-states obviously increase the scale of wars, the cumulative war mortality of the entire population is twenty times less than for tribal societies, although this ratio would change in the event of a full-scale nuclear confrontation.

Anthropologists, historians, and especially philosophers have often underestimated the ferociousness of life in preindustrial societies, nonetheless. This reflects romantic ideas about the essential goodness of human beings in a state of nature. These ideas were articulated by influential French philosopher Jean-Jacques Rousseau, who argued that people are corrupted by civilization and ennobled by contact with the natural environment. These ideas have had a seductive appeal for writers for more than two centuries, but as far as violence is concerned, they are clearly wrong. Economic development is generally accompanied by increased restraint of military violence.

In addition to assuming that primitive people were more pacific than the data indicate, historians and anthropologists often underestimated the military technology of so-called primitive people. Archaeologists find otherwise, however. In one thousand-year-old society in San Francisco, 5 percent of the skeletons recovered in an excavation still had obsidian arrowheads sticking in them. This is the most obvious sign of death in war but it clearly underestimates the fatalities. Approximately one-quarter of the people at this site died from war injuries, according to informed estimates.[12]

Were these early San Fransiscans unusually belligerent? No. If we judge from extant tribal groups, as well as other archaeological evidence, they must have been about average. Among the Jivaro of Peru, who practice headhunting, over 30 percent of people die violent deaths. Among the Qadan of ancient Egypt, the number was over 40 percent. To put these numbers in context, the likelihood of dying in warfare in economically developed countries is only around 1 percent, at least so far, despite the spectacular scale of modern wars and the frightening power and efficiency of modern military technology. Mortality is much higher for males in the preindustrial groups, with 50 percent of Qadan men and 60 percent of Jivaro men dying in battle.

Why do tribal wars produce such high mortality rates? The most important reason is that battles are so frequent. Among warlike societies, such as the Yanomamo of South America, skirmishes between hostile villages are an almost daily occurrence. Such battles usually do relatively little damage but have a cumulative effect: over one-quarter of the men die from battle injuries.

Why are local battles so frequent and chronic in some societies? The fundamental problem is that once hostilities have been launched, it can be very difficult to make peace. Given that all neighboring villages are in competition with each other over vital resources, like land and women, there is no genuinely neutral third party. If villages A and B are tired of their chronic hostilities and want to make peace, A could theoretically go to village C and ask headman C to approach B to make peace. This is dangerous, however, because C might go to B and say, "Village A is weak; let's join our forces and finish them off."

Village-level wars may sometimes continue until one group completely wipes out all the warriors in the other one. In developed countries, military procedures recognize a right of surrender. By surrendering, millions of soldiers in defeated nation-states have saved their lives, and such clemency helps to salve the bitterness of defeat. Surrender is typically not an option in village-level, or tribal battles. If you spare an enemy, you can be sure that he will return to fight you another day, making you regret not pressing your advantage and killing him the first time. It is true that prisoners were sometimes taken in preindustrial wars, but being taken prisoner was a virtual sentence of death that was often preceded by grotesque torture. When it came to war, the noble savage was often more savage than noble.[13]

The lack of mercy in battle is not just a function of inadequate political structures for making peace. The whole reward structure within warlike societies favors homicidal aggression. Individuals who are reluctant to fight are not commended for their humanitarian restraint. Instead they are ridiculed as cowards and lose social status. Among the Yanomamo, warriors who have killed an enemy are referred to as *Unokai* and enjoy respect and prestige. They are attractive to women and thus translate their status into increased reproductive success. Unokai have more children than average, according to Napoleon Chagnon's research.[14] If warlike risk-taking has an immediate payoff in terms of access to women, this can outweigh the fitness costs of dying young in battle.

Given the nastiness of preindustrial warfare and the rewards heaped on homicidal warriors, it might seem difficult to make the case that humans are fundamentally altruistic creatures with strong dispositions to help and support others in their social group. Yet, this is not really such a contradiction because aggression against enemies involves personal risk and sacrifice and may be motivated by altruistic feelings toward one's own social group. As societies have become more complex, they have devised successful methods of tapping in to this visceral altruism toward family and tribe.

IN-GROUP ALTRUISM IS OUT-GROUP ATROCITY

Atrocities have always been committed against out-groups. English Puritan Oliver Cromwell had no trouble carrying out grisly massacres against Irish Catholic men, women, and children. During the African colonization, British troops were happy to mow down spear-clutching Zulus by the thousands, using machine guns. No world power has ever been immune from this kind of bias in its treatment of out-groups. Even the United States is guilty of atrocities committed by renegade officers against Vietnamese villages that involved the murder of babies predicated on a complete dehumanization of the enemy.

Cynics are inclined to view such episodes as characteristic of human depravity and to claim that we are therefore incapable of altruistic conduct. Yet, it is a grave logical error to use these exceptional events to condemn our everyday behavior. Atrocities are committed against out-groups, but they can be motivated by a very strong altruistic identification with the in-group. That is an irreducible paradox of human group behavior.

To say that atrocities are an almost inevitable consequence of human organization into antagonistic groups may seem like a spineless evasion to many moralists, rather like Eichmann's defense that he was just obeying orders. The practical question to ask is,

"How can you tap the altruism that undeniably exists within groups to prevent aggression against out-groups?"

Luckily, the answer to this question must be reasonably simple, because the story of recent history has been one of increasing complexity of social organization. This means that the altruism within families has been tapped by local communities. The altruism within communities has been tapped by tribes. Finally, the altruism within tribes, or ethnic groups, has been harnessed by nation-states.

These processes crop up in the history of all countries and should therefore be quite easy to investigate. In many underdeveloped countries of Africa and Asia, central governments are still comparatively weak. They have trouble collecting taxes, which are often interpreted as little more than the theft of assets by a politically dominant clan from weaker ethnic groups. With lack of resources and the squandering of public funds through corruption, the governments in these countries accomplish little, as measured in terms of infrastructure, public education, public health, national security, law enforcement, economic development, and so on. Apart from feathering their own nests, governments do little but attempt to adjudicate the excesses of local conflicts between rival ethnic groups organized under tribal leaders or warlords.

All effective modern nation-states went through a process of unification of tribes, sometimes through negotiated treaties, and sometimes through armed conflict. The dominant faction ultimately imposed its will on weaker ones, resulting in a unification of languages, customs, laws, and administrative structures. In many modern countries, like the United States, England, and Germany, the process of integration has gone one step further as immigrants from various countries enter a melting pot, giving up their allegiance to a former country to attain citizenship in their adopted nation.

The success of the melting pot may be qualified, at least in early generations. Thus, US immigrants from Ireland cannot have been too happy to read signs saying, "No Irish need apply," and many Turkish immigrants to Germany have had similar experiences in the form of hostility from neo-Nazis. Almost inevitably, however, immigrants trade some of their original identity as they become fully integrated after about two or three generations.

The fact that people's group affiliations are so plastic might suggest that the answer to international conflicts is a world state, a gigantic melting pot in which all individuals take out world citizenship. From a practical perspective, it does not seem like a very good idea, however, because a single world state would likely be an appalling tyranny. The most conspicuous cases of ideologies that attempted to establish world dominion have not been encouraging. They include Nazism under Hitler and Communism under the Soviet Union.

World religions have also been an important unifying force for millions, or even billions, of individuals from various countries around the world. In general, religions do not have administrative missions, however. Even if they did, a world government, as a theocracy, is hardly more reassuring in respect to tolerance and individual freedom than Communism or Nazism.

If a world government is not a good idea, how can countries solve the problems posed by international conflicts of interest? How can they ever hope to address global problems, like atmospheric pollution, nuclear proliferation, and terrorism? It should be clear that these problems are the same, in principle, as the tragedy of the commons where individual selfishness degraded public property. This simple comparison has helped game theorists to propose effective solutions to these pressing global problems. Before getting to this point, though, it is essential to understand some of the ways that altruism is tapped within countries despite conflicts between the duties of a citizen and individual freedoms.

THE TIES THAT BIND A COUNTRY

How does a nation emerge from the conflicting tribal groups from which it is commonly derived? In some ways, the process is just an extrapolation of the formation of tribes from kin groups. In taking on tribal obligations, such as military service at the command of a chieftain, families clearly sacrificed resources and individuals that would otherwise have contributed to the welfare of kin groups. Why make such a sacrifice? Clearly, the payoffs of incorporation into larger entities, such as protection from local attacks, had to exceed the costs.

Military security might not be the only issue, of course. Thus early urban civilizations sprang up around large water irrigation projects in the Middle East. The cities of Mesopotamia often had as many as 100,000 people living in the same economic system organized around a public irrigation system that provided the water for a highly productive cultivation of cereal crops. Irrigation systems were built, and maintained, through public works programs without which agricultural productivity would have been considerably lower. It is thus fairly obvious that even after obligatory contributions of labor to the ambitious public works projects, and public monuments that were undertaken largely as an ego trip on the part of rulers, workers were better off in these early "hydraulic" societies than they would have been if they had scratched out independent farms on the banks of some remote stream. They were certainly not affluent, wealth being sucked up the social hierarchy, but they were assured a steady supply of food. Wise rulers also built secure granaries to store grain that could be released to the public during periodic crop failures and famines, as illustrated in the biblical story of Joseph, who administered the grain surplus of the pharaoh.[15]

A similar rationality can be discerned in the formation of modern nation-states in Europe to increase military security. Prior to the fourteenth century, Europe was organized under a feudal system according to which the local lord was monarch of all that he surveyed but swore allegiance to a more powerful regional figure who was responsible for assembling armies in the event of military invasions. These liege lords swore fealty to a national monarch who was often little more than a figurehead and exercised little real authority throughout his nation, despite having many responsibilities, particularly in respect to national defense. The rise of powerful nation-states in Europe was partly a product of internal politics and partly a function of developing military technology. The internal force was the rise of cities as economic powerhouses. In order to weaken the influence of the local squire, cities frequently threw in their lot with the monarch. In the process, they acquired administrative independence and avoided paying taxes to the local lord.

The origin of modern states owes a great deal to military technology, however. Advances in siege warfare meant that only centralized states could bear the great economic burden of defense. With increasing power of the state, feudal lords were forced to give up much of their authority in return for security against invasion. There is thus a parallel to the development of tribal societies, in which clans throw in their lot together to fight off rival tribes.[16]

By the end of the Crusades, the holy wars fought by Christian knights against the Byzantine and Islamic empires, in 1270, munitions had begun to play a minor role in warfare. In 1249, the Crusaders were attacked with primitive bombs made of gunpowder that were not very destructive but made a frightening noise on the battlefield. Early in the fourteenth century, cannons were first made in Europe but they did not play an important role in battle until the middle of the fifteenth century. Muskets appeared in the sixteenth century and were followed by pistols. Armed infantry soon overshadowed earlier types of warfare, including swordsmen, pikemen, archers, and even mounted knights that had previously plowed through the lines of infantry, like modern-day tanks.

With the development of effective weapons for infantry, foreign armies were more threatening, and the seventeenth century saw a great increase in the size of European standing armies. The threat was magnified by development of increasingly powerful, and increasingly accurate, cannons that were capable of reducing previously impregnable medieval fortresses to a pile of rubble within a few hours.[17]

The only defense against siege cannons was to build larger, more elaborate fortifications that contained many defensive guns, which were designed to blast away at the gun emplacements of attacking forces. European governments invested huge amounts of money into defending their cities and ports from invading forces. The great expense could be met only by increased taxation. Centralized governments levied taxes on citizens to meet their military budgets. With increased centralization of taxation came an increase in the power of centralized governments. Medieval Europe thus countered the military threat of invasion with an increased centralization of government. In hydraulic societies, by contrast, centralized government emerged much earlier in the history of cities because it organized the construction of irrigation systems that were the key to subsistence.

Just as farmers in Mesopotamia banded together into towns and cities to benefit from the economic advantage of a public irrigation system, so the feudal lords were obliged to accept centralized authority in the name of military security. The combination of political units into ever-larger entities is not inevitable but is dictated by a calculus of changing costs and benefits. In this sense, the centralization of governmental authority in Europe was just as rational as the centralization of power in Mesopotamian cities. In each case, the loss of autonomy was compensated for by an increase in the scale of political altruism and a corresponding enhancement of the common good.

Larger political groupings can thus provide practical benefits for their members. This is particularly obvious where large-scale cooperation is necessary for subsistence, or for defense against enemies. Just as there are many situations when defection is in the interest of the individual, as illustrated by the prisoner's dilemma, so there are some situations in which cooperation with groups has overwhelming advantages. In self-sufficient island communities, for example, a high level of cooperation is necessary because there is so much mutual dependence.

Larger social groups may provide practical benefits for their members. Yet, that does not explain how the psychological processes of identification with small local groups, whose members are all well known, and frequently encountered, can be applied to much larger, and more anonymous, social entities.

Social psychological studies of group formation are very helpful in this respect. Even with minimal exposure to a group that they know to be arbitrarily formed, people behave altruistically on behalf of groups with which they are identified (as discussed in chapter 9). There could hardly be a more compelling illustration of our groupish tendencies. Given the human capacity for identification with various social groups with which we have no specific history, it is not surprising that we should be capable of forming allegiances to large social units of various kinds, whether they are tribes, nation-states, or empires. Anthropologists describe numerous social organizations to which people belong in subsistence societies, including peer groups of children and adolescents, men's houses, religious groups, foraging units and other work groups, military units, and so on.

In economically developed societies, there is a far greater range of organizations that people may join, and many of these organizations have great size and influence. The largest global organizations are religions because the most successful religions are good at recruiting converts. Religions can be thought of as large mutualistic organizations, analogous to secular states. Despite some highly conspicuous examples to the contrary, such as sexually abusive priests, it appears that people join religions because of the individual benefits of being part of a large mutualistic organization.

THE POTENTIAL BENEFITS, AND COSTS, OF RELIGION

Some religions are allied with states. The Church of England, for example, was the official church and assessed tithes, or taxes, which made it enormously wealthy and powerful. With the rise of powerful centralized states, the power of taxation shifted to governments. Many of the states of the Middle East are described as Islamic republics, which means that the goals of church and state are still, at least theoretically, the same. In general, religion does not do well when it is actively opposed by the secular state, as happened in the Soviet Union, where religious observance was repressed. Most influential world religions minimize conflict with secular authority. They render unto Caesar that which is Caesar's and render unto God that which is God's.

Religions are particularly interesting to students of altruism because of the self-sacrificial fervor that they may incite in members. Religious people feel differently about the world and they may be motivated to act in unusual ways, such as making great sacrifices to go on a religious pilgrimage, going abroad to do missionary work, or dispensing all of their worldly property to the needy. These behavioral dimensions of religion are highly variable even within the same religious community. In some cases the flow of altruism is in the opposite direction, with people joining religions because of the worldly benefits they may obtain, such as inexpensive education. A Russian emigre confided to the author that he had joined a religious community in Moscow and attended Sunday services purely for the coffee with cakes that was served afterward. Similarly, street bums may attend mission services because of the nourishing soup that is provided afterward. Such cupboard religion may be rare, however, and confined to people who find it very difficult to make ends meet.

Although religious organizations often have very complex relationships with their communities, the benefits of religious membership must generally exceed the costs. Some of these potential benefits are quite subtle, or nonmaterial. Perhaps the most important advantage for members of a religious community is an increased sense of social integration. People who have little other forms of social support but attend religious ceremonies may feel that their beliefs, and opinions, are shared by a large number of similar others and they may believe that their lives are part of some divine plan. Those who cannot participate in such shared experiences are more likely to feel socially isolated, or lonely. In the language of sociologists, they suffer from *anomie,* or social alienation. Religion may provide people with a sense of purpose in life, although there are several other nonreligious sources of mission, groups with highly ethical concerns, such as those of secular humanism, global development, and ethical philosophy. Some people dedicate their lives to science, or the arts. Others are completely absorbed by money, politics, or sports.

As already suggested, religious communities can provide important material benefits, such as inexpensive education, and, for businesspeople, exposure to members of the local community who are potential customers or business partners. As in the case of volunteer service in charitable organizations, church attendance could be maintained by such ulterior motives.

The primary benefits of religion are mostly psychological, however. Jesse Ventura, the governor of Minnesota, alienated his electorate by expressing this idea in a disparaging way when he said that organized religion was a crutch for the weak-minded.[18] Whether religious people today are weak-minded or not is debatable. Throughout the broad sweep of human history, the majority have been religious, adopting the same creed as their forefathers. In the past, the most brilliant people were just as religious as everyone else. Today, atheism is growing rapidly among educated people. The majority of scientists (60 percent) are atheists, for example.[19] Among elite scientists, the probability of being a nonbeliever is much higher.

Perhaps the most compelling argument in favor of the psychological benefits of religion is the possibility that religious people are healthier. We know that social integration and the feeling that we live in a controllable world (the opposite of psychological stress) are predictive of better health. Does religious observance actually improve health? Does it improve reproductive fitness?

RELIGION AND HEALTH

The fact that religion crops up in some form in every society studied by anthropologists suggests that it likely provides significant benefits for health and survival. Religion may either help worshipers to solve the practical problems of finding food and shelter, or help some at a psychological level reduce chronic stress that can devastate health by aggravating heart disease and asthma, or even degrade the capacity of the immune system to ward off infections.

Despite the generalization that organized religion must be beneficial for believers, some religions are obviously very bad for the health of their members or even for the religion's own survival. At an extreme, cults like Heaven's Gate and Jonestown organized mass suicides of their entire congregations. These religious movements now appear to be defunct. The Shakers, who renounced sexuality, will not have much impact in the future, either, because they failed to produce children to replenish the numbers of their congregations, relying entirely on recruitment to gain new members. Although Shaker communities were economically, and socially, successful in their day, they are currently reduced to a handful of geriatric survivors living in Maine. Viewed as an evolutionary experiment, the Shaker religious system has been consigned to the scrap heap of history. It has been weeded out by natural selection, even though Shakerism has no genetic basis, as far as we know.

Such religious communities provide compelling examples of why religions that do not further the biological interests of their members go extinct themselves. It is not much of a stretch to suggest that established religions got that way by furthering the reproductive interests of their members. This seems obvious in the case of pronatalist religions like Catholicism, which rejects effective forms of scientific birth control, thereby ensuring a steady growth in membership. Islamic women often have large families, as well, so that religious membership enjoys a rapid rate of growth.

Religions that promote reproduction will clearly increase in numbers and influence. One way that religion promotes reproduction is by encouraging marriage. Young people with shared religious beliefs are more likely to meet and marry, for example. Exposure to each other each week at religious ceremonies provides ample opportunity for the sparking of romantic interest, if nothing else. Religion is important in the selection of a spouse: couples are very likely to be of the same religion. Increasing the reproductive success of members is one way in which religions can contribute to the biological success of members, and thus to the success of the religion itself. Another way in which religions can increase the biological fitness of members is by promoting their health, whether through psychological benefits or through a healthy lifestyle. A great deal of research has recently addressed the question of whether religious observance improves health.

Even if a religion claims to promote health, as Christian Scientists do, it can have the opposite effect.[20] In reality, members have unusually bad health and do not live long lives. For Christian Scientists, healing is seen as a spiritual process, rather than something that can be facilitated through scientific medicine. They are subject to illnesses without recourse to medicines or surgery. In an offshoot of Christian Scientists, the Faith Assembly, members often avoid all medical care and fare even worse. Without going to this extreme, many religious groups raise objections to specific life-saving medical interventions, including blood transfusion, organ transplantation, and immunization, usually on the basis of some concept of bodily purity. In renouncing immunization against diseases like measles and polio, some groups suffer little in terms of increased illness because they are shielded by living in a larger society where the majority are immunized, making it difficult for epidemics to get started. Transplantation operations still affect only a small minority of the population, but refusing blood transfusions can increase mortality significantly because about half of the US population receives blood at some point in their lives.

Despite religious objections to some medical procedures, research suggests that regular church attendance and daily prayer is associated with better health, although this is a matter of ongoing scientific controversy.[21] Some research has shown that religious people have lower blood pressure, suffer fewer strokes, have more robust immune systems, and experience less depression. What is more, these health advantages can translate into large differences in lifespan. One expert concluded that the increased mortality rate associated with a lack of religious involvement is equivalent to the effects of smoking a daily pack of cigarettes for forty years.[22]

A large-scale study of the elderly in Piedmont, North Carolina, reported that those who were not regular attenders of religious services had a 63 percent increased mortality rate.[23] Another recent nationally representative study, by Robert Hummer of the University of Texas at Austin, and others, found that frequent church attendees lived seven years longer than those who never attended, although most research finds more modest effects, or none at all.[24] It thus appears that there are differences in health between regular religious observers and others. On this issue there is little controversy. The real problem is in accounting for such differences.

If the content of religious belief, or specific religious practices, or lifestyle influences were important, we would expect to find telling differences among different denominations. In reality, there is no great difference in health outcomes between various mainstream religions. The major branches of Christianity and Judaism enjoy similar health advantages, for example. There has been little investigation of Islam, Buddhism, or Confucianism, the other major world religions, or others who regularly meet in a group.[25]

Why might religious people be healthier? Most researchers rule out supernatural influences, such as the power of prayer by others on a person's behalf. Interestingly, daily prayer can improve health for the prayer himself or herself, an advantage that is indistinguishable from secular meditation. Health researchers are dedicated to the finding of natural explanations, rather than supernatural ones, of course. They assume that the health benefits of religious attendance are somehow derived from lifestyle.

Membership in a religious congregation increases social support, which has critical importance for health and longevity, as already pointed out. Church members notice if someone is absent and ensure that they receive medical attention when needed. Religious communities may also serve as loose friendship networks. They are thus a source of emotional support that speeds recovery from illness.

Statistical control for social support helps to explain some of the health advantages of religious people. Even with social support statistically factored out, religious people still have a health advantage in many studies. This is not simply a question of healthy people being more religiously active because they are more energetic. When matched populations are studied over time, religious people tend to improve in health compared to those who are not religiously active.[26]

Researchers assumed that this benefit was due to the healthy lifestyle promoted by most religions. Thus, many religions discourage, or prohibit, the use of alcohol, tobacco, and other recreational drugs. Many discourage gluttony and sexual promiscuity, which can undermine health in various ways. Gluttons are likely to be overweight and have greater vulnerability to major killers like heart disease, stroke, cancers, and secondary diabetes. Promiscuous people have greater exposure to infection from venereal diseases, which may undermine reproductive function, as in the case of chlamydia for women, or cause early death, as in the case of HIV/AIDS today, or syphilis in the days before antibiotics.

According to Aldous Huxley, the tragedy of science is the killing of a beautiful theory by an ugly fact. Various religions differ greatly in recommended lifestyles. Yet, the lifestyle differences are not obviously predictive of health. Clean-living Seventh Day Adventists are not appreciably healthier than the more libertarian Episcopalians, for example. This ugly fact calls the lifestyle explanation into question. Formal research shows that when lifestyle factors are statistically removed, religious people, as noted, still have a substantial advantage in health and longevity. We are beginning to understand why.[27]

Researchers at Duke University have demonstrated that individual prayer, as well as attending church services, reduces blood pressure. Religious practice also helps people to minimize stressful life events. Probably for this reason, religious people are less prone to depression and bounce back more rapidly from a depressive episode. Other interesting objective evidence includes the fact that their immune systems are more active, producing larger quantities of interleukin 6, for example. This suggests reduced exposure to stress hormones that impair immune function, which, in turn, supports the conclusion that religious people may be better at handling psychological stress. Most religions promote the view that human life is worthwhile and has a purpose. There is abundant evidence that such an optimistic frame of mind promotes happiness and health.[28]

While most researchers agree that religiously active people are healthier, they are also of the opinion that the health advantages derive from social support, and other lifestyle factors and have little or nothing to do with the content of religious faith, a point that is borne out by the very poor health of those religious communities that depend on faith healing as opposed to scientific medicine. If religious belief is associated with improved immune function and better health, it does not follow that religious belief is necessary to achieve these benefits. A weekly poetry reading might conceivably have the same health benefits as a religious ceremony. In fact, the weekly services of Unitarians often include secular poetry readings and musical recitals, for example. As mentioned, the secular use of meditation, a technique that derives from Eastern religious practice, is effective in reducing blood pressure and thereby helps people to manage life stresses. What is more, there is no evidence that atheism, of itself, undermines health. Arch atheists George Bernard Shaw and Bertrand Russell lived to the ripe old ages of ninety-four and ninety-eight, respectively. These tough-minded individualists probably had more purpose to their lives than most religious individuals.

The whole area of possible religious effects on health has been rife with academic controversy. Two review papers concluded that there were unmistakable signs of the health advantages of religion, so much so that religion could be viewed as a form of alternative medicine. Inspired by such claims, most medical schools in the United States now have religion courses as part of their curricula. Nonetheless, in 2002, Richard P. Sloan and Emilia Bagiella of Columbia University published what could be described as a blistering attack on the earlier reviews of the literature.[29] They argued that none of twelve studies in the first review (by F. Luskin) provided a sound scientific basis for concluding that religion improves health. In the second review paper (published in the *Handbook of Religion and Health*), only four of the thirty-nine studies cited provided convincing evidence for an effect of religious observance on health. In their own review that focused on all the 266 English-language abstracts on religion included in Medline for the year 2000, only 17 percent provided evidence relevant to potential health benefits of religion, suggesting that much of this literature is not helpful in deciding whether religion affects health.

Despite the handful of studies with positive findings, Sloan and Bagiella concluded that there is little scientific support for the hypothesis that religion improves health. This would appear to be a swing of the pendulum from uncritical acceptance of religious effects to the opposite extreme of denying any positive findings in their own review.

To return to the main questions, it is clear that major religions can promote healthier lifestyles as well as increasing the reproductive success of members. This helps to explain why a propensity for religious belief could have slipped through the net of natural selection, even though it can promote irrational beliefs and practices. Religions are an interesting example of one way in which large self-sustaining organizations can draw on the capacity of individual members to behave altruistically toward other members of their groups.

In theory, religions could serve as one way of uniting the nations of the world into harmonious communities, and this has certainly happened in the case of the Roman Empire (in its Christian period) and the Ottoman Empire that followed. In practice, in-group altruism has meant out-group aggression for religions as well as for other social groups, like tribes. Many religions treat out-groups as inherently subhuman. They use belittling terms such as heathen and infidel. Human history is punctuated by religious wars, like the wars of the ancient Israelites against their numerous idolatrous enemies, the Roman war against early Christians, the religiously inspired second-century-BCE uprising of the Maccabees against the Seleucids (Hellenistic rulers of Syria/Palestine), the Crusades fought by European Christian knights against anyone they encountered on the way to Jerusalem, and the current terrorist war fought by Islamic extremists against everyone else.

Religious wars generally have very little to do with theological conflict, as such, but religion is used as an identifying tag in wars between ethnic, or national, groups. Such conflicts are often predicated on conflicts over vital resources, like land, oil, or employment opportunities. Thus, the Palestinians and Israelis are split on religious lines, but their real struggle is over land and sovereignty. In Northern Ireland, Irish Catholics rebelled against the Protestant regime because they were poor and encountered discrimination in housing and jobs analogous to discrimination against African Americans in the United States before the civil rights movement.

Even if religious groups discriminate against members of other religions, one would expect them to support those within their group. There is plenty of evidence that indeed shows that they do favor those of their sect with friendship and charitable donations to their own denomination. Despite a widespread perception that religious people should behave more ethically in general, and toward fellow members of their sect in particular, there is surprisingly little evidence that religious people either reason with a higher level of ethical sophistication, or avoid dishonesty in dealings with their sect and others, although they do have an advantage in terms of lower criminality.[30]

Observers are often struck by the high level of religious faith in jails, however, even among the most violent and depraved of offenders. To some extent this phenomenon may be a side effect of the "misfortune" of criminal conviction that prompts prisoners to reflect on the error of their ways. This incentive is particularly strong in the case of the small minority of prisoners who receive death penalties. Religious conversion in such cases may be undertaken as a sort of "plea bargain" according to which they expect to receive more lenient treatment in the next life for their misdeeds.

Researchers who examine the influence of religion on ethical behavior of ordinary people find that there is a baffling inconsistency in their results, suggesting that religious people are not generally more ethical than others and may behave less ethically than atheists. Thus, one study, published in 1975, found that atheists were significantly less likely than religious students to cheat on an exam. This is not a trivial matter in a country where some three-quarters of students admitted to academic dishonesty. Given that some 90 percent of the population were religious believers, this suggests that the majority of religious students were academic cheats.[31]

Students of moral development are inclined to conclude that religious belief stunts moral development, because it commits people to a dogma, or formula, rather than working out ethical solutions for themselves, which is considered to be the highest stage of moral development (postconventional morality). People who adhere to an established religion are thus less likely to attain the highest levels of moral reasoning.

Religious texts may include exemplary tales that endorse a low level of moral reasoning. Thus, in the biblical story of David and Goliath, David is not content with humiliating the enemy warrior by rendering him unconscious with the impact of a stone to the head but takes the opportunity to cut off his head using the giant's own heavy sword. Following this brutal deed, he becomes a hero of the Old Testament. There are numerous morals that can be drawn from this story. Here are some of the more obvious:

- Skill is more important than strength.
- Always press home a military advantage and give the enemy no quarter.
- Violence solves problems.
- If you are good at killing enemies, you may become popular.

Each of these morals encourages people to operate at a low level of moral reasoning. From that perspective, it is fascinating that Christians continue to find inspiration in the Old Testament with its endless barbarity despite preaching a message of peace derived from the New Testament. A superficial survey of Christian history makes it clear that belligerence there too has often won out over peace-making.

Religious people often claim the high ground in making the case that more religion will help us to fight crime, save the family, cure drug addictions, protect children, and even avert wars. Living in a deeply religious country, we are so accustomed to hearing these claims repeated that it is something of a shock to discover that there is little compelling evidence that religious people are more ethical, or actually live better lives than the nonreligious.

Some scholars have even made a case that fundamentalist religions undermine moral reasoning. For people who "know" that they are saved, actions may be irrelevant to their fate in the afterlife, so that they are relatively unconcerned about who is hurt by their actions in this world. It has even been argued that religiosity may predispose people to sexual molestation of children.[32]

Knowing about a person's beliefs, or the fervor of his religion, tells us virtually nothing about his ethical behavior to the bemusement of sociologists and psychologists, who have studied this problem over many decades. Some expected that deeply religious people would behave more charitably because sacred texts in Christianity and other world religions exhort believers to behave charitably and with compassion, nonetheless social scientists could find no evidence of this. Among the findings of their research, as reviewed by sociologist Alfie Kohn,[33] were the following:

- A 1950 study of Episcopalians found no relationship between involvement in religious activities and charitable acts.
- A 1960 questionnaire study found that belief in God was only slightly related to altruism. Attendance at religious services was completely unrelated to altruism.
- In 1984, interviews with seven hundred residents of a medium-sized city found that religious people were not particularly good citizens, as determined by involvement with neighbors and participation in local organizations.
- People who rescued Jews in Nazi Germany were not any more religious than nonrescuers.

The only reliable difference between religious people and others, according to the research literature, is an unflattering one: religious people are more intolerant of ethnic minorities.

Despite religious warfare, and the surprising lack of influence of religious belief on ethical behavior, religion can play a constructive role in uniting countries. This is close to the surface in the United States, where presidents never give a speech without mentioning God, thereby appealing to a majority of the electorate. Despite the formal separation of church and state in the constitution, religious faith is pushed in the currency, in the pledge of allegiance to the flag recited daily in schools, and in the national anthem, sung at all major public events. Anyone who gives evidence in a court of law must implicitly acknowledge that they are Christians by swearing on the Bible, a ceremonial acknowledgment that what they are saying in testimony is as true as the content of the holy book. For atheists and people of other religions, swearing on the Bible thus provides little guarantee of veracity.

Such rituals may be performed as much from historical habit as anything else, but it is quite clear that, in the past, religions were a powerful cohesive force in states. Arguably, the role of religion as a kind of social glue is giving way to cooperation based on economic transactions that are mediated through money.

MONEY AS RECIPROCAL ALTRUISM

In modern economies, most of our social relationships, outside family and friends, are mediated by money. All the goods consumed in a home, and all of the outside services necessary to make it run, are purchased. We spend most of our active days at work building relationships with people we would not choose to know if they were not cogs in the economic machine that generates our income.

To many people, connecting economic relationships with altruism may seem forced. For them, as for economists since the days of Scotsman Adam Smith (1723–1790), a key founder of the discipline, people are intrinsically selfish, and the pursuit of their self-interest, in seeking to maximize profits, has the happy consequence of improving economic efficiency through competition, and thus of contributing to the wealth of the entire community.

We often think of economic transactions as potentially selfish, cold, and calculating, because they supplant an earlier system of altruistic interactions based on a combination of kin-based altruism and mutualistic relationships among people who are well known to each other, like a group of !Kung hunters. If you own a store on a busy city street, most of your customers will be strangers. Yet, by shopping with you, these strangers help to pay your rent, feed your children, and allow you to shop at other stores to purchase whatever you need. Despite its cold impersonal feel, the monetary economy is a system of reciprocal altruism in which goods and services given out today are reciprocated later through the medium of money.

Monetary economies, like religions, have the capacity to bind complete strangers into extended reciprocity networks. These networks not only bind nations together into effective societies, but also extend around the world, tying nations together through trade and commerce. The phrase "global village" is often bandied about in terms of electronic commerce, but this grandiose concept is quickly becoming a reality. A man in New Delhi who orders flowers to be sent to the home of his daughter on the birth of her first child in Little Rock, Arkansas, for example, is a participant in the Arkansan economy just as much as the daughter who lives there. With a monetary economy, and electronic communication in business, every person in the world is potentially linked into a reciprocity network with every other person.

To many, the globalized economy is a frightening prospect because it undermines the local autonomy of farms, industries, and governments, all of which become very small fish in the very large pool of international commerce. Such ongoing economic dislocation, nonetheless, has been a feature of life since the Industrial Revolution. The most prominent form of dislocation was the literal movement of people from farms to cities in search of employment. Urbanization can bring initial hardships but it is the major source of national wealth. Thus, the poorest countries of the world are the ones, many located in Africa, that retain a large rural population. Many rural Africans practice subsistence agriculture that provides for their immediate needs but contributes little to the national economy. Improvement of their standard of living, education, and health can be achieved only through urbanization because that is the only way that countries can grow their economies substantially and generate sufficient tax revenues to provide funding for public services. When people work in cities, they generate income that both stimulates the local economy when it is spent, and can easily be taxed by governments. Prosperous businesses are also a major source of government revenues. For these reasons, developed economies can provide humanitarian services to the needy, further illustrating how wealth and altruism can go together.

Urbanization may increase national, and global, altruism, but it is also largely responsible for the problem of global pollution and other ills of civilization, such as infectious disease epidemics, however.[34] Extreme environmentalists suggest that urban development has been a disastrous mistake, whereas subsistence agriculture is ecologically sustainable. Whether they are correct or not, it seems unlikely that any nation will voluntarily renounce the benefits of industrial development in favor of a Gandhi-like simplicity of living that emphasizes self-sufficiency of the local economy and suspends urban centralization. Given the equation of industrial development with pollution, can human altruism save the global environment?

CHAPTER 11

Saving the World

A globalized economy encounters serious problems, ranging from disease epidemics, like HIV/AIDS, to international terrorism and environmental pollution. How can the evolved altruistic tendencies in each of us be tapped to address such monumental problems? Environmental problems offer useful insights.

SAVING THE ENVIRONMENT IS A START

We are well on the way to becoming an urban species that makes, and consumes, an increasing volume of products, necessitating huge expenditures of energy as well as generating vast amounts of trash. This is a profound ecological change compared to our ancestors, who made their way in small hunter-gatherer groups before gravitating to larger horticultural, and agricultural, villages. Human urbanization resembles the development of large colonies of social insects. Both involve control of the microhabitat through construction and temperature regulation, and specialization of work roles.

Evolution has had no time to change us for our new urban way of life. Yet, urbanites are unlikely to return voluntarily to subsistence farming, much less hunter-gathering, unless they are forced to do so by extreme environmental degradation that makes cities unlivable or threatens the viability of the global atmosphere.

Rumors of the demise of the global atmosphere due to pollution may have been exaggerated, and this is illustrated by the issue of carbon emissions and global warming. These problems are real, of course, but their impact on the environment is complex. Thus, environmentalists have long assumed that the amount of carbon in the atmosphere is a direct function of carbon dioxide pollution due to human activity. It has recently emerged that some regions of the globe, specifically North America, suck carbon dioxide out of the atmosphere. Hence the term *carbon sink*.[1]

There is no doubt that heavy industry has always caused serious air pollution (see fig. 12) and respiratory diseases because the discharge of contaminants exceeds the capacity of the environment to absorb them. The human respiratory system has not been designed by natural selection to cope with large amounts of fine dust characteristic of industrial pollution. The decline in such pollution from heavy industry in some affluent countries coincides with its increase in developing economies, as dirty industries get exported from wealthy countries to poorer ones due to complex influences of global commerce, including the cheapness of labor in developing nations. The export of polluting heavy industries to developing nations greatly complicates international efforts to protect the world's atmosphere and other degradable natural resources, particularly bodies of water and fragile ecologies such as rainforests and coral reefs. The tragedy of the commons thus plays itself out on a global scale.

Global pollution can be thought of in terms of a prisoner's dilemma game. Polluting nations benefit their own economies when they consume energy, produce goods, and thereby pollute the environment. They are like the selfish herdsman putting out another sheep on the overgrazed commons. From one perspective, nations are not really voluntary players because they have little choice about whether to be economically productive or return to the Stone Age. All nations strive to be economically successful, just as herdsmen struggled to make a living from the commons. In that sense, all countries participate in the tragedy of the commons, that is, of environmental degradation by industry.

Fig. 12. Environmental pollution pits the interests of businesses against communities and of individual states against the global environment. (Library of Congress)

This is a real and pressing problem but it does not have to be an unstoppable tragedy. We know that environmental problems can be solved through concerted action of leading world nations. Thus, the hole in the ozone layer of the atmosphere used to be a major threat. Created largely by chlorofluorocarbons (CFCs, and also methyl chloroform and carbon tetrachloride) used as aerosol propellants and in air conditioning systems, thinning of the ozone layer is a serious problem because it exposes humans to high levels of dangerous ultraviolet radiation that causes skin cancers. The ozone hole was first observed over Antarctica in 1985, and there has been a worldwide increase in skin cancers since then. The good news is that in Montreal in 1987, the United States and twenty-two other countries agreed to phase out CFC production. In the final treaty, 107 countries agreed to ban CFCs by 1996. Rapid cooperation leading to action thus protected the commons of the ozone layer from further deterioration. It is as though herdsmen noticed damage to the pasture and voluntarily agreed to limit grazing.[2] Contrary to the received wisdom in economics, tragedies of the commons *can* be averted, if only the participants get together and agree on behaviors that are to their mutual long-term benefit. This logic applied to states also works for individuals, of course, as when people agree to pay taxes to support local services like trash collection.[3]

As any environmentalist can tell you, responses of the world community of nations to environmental problems has not always been either so prompt, or so decisive. The issue of global warming and the failure of the 1997 Kyoto agreement, designed to limit emission of greenhouse gases, is an instructive example. Kyoto failed because its architects made the inadvertent mistake of pitting the self-interest of developing nations against that of wealthy countries, thus undermining its principal goal, namely, to improve the global environment for the benefit of all countries.

Architects of the Kyoto agreement were aware that pollution control has an economic price and that the expense of pollution-control measures could cripple the economies of developing countries. How could one justify increasing the economic miseries of people in developing nations? Their solution was to excuse developing countries from their contributions to the greenhouse effect and to assume that regulating greenhouse gases in wealthy countries, like the United States, that consume the great bulk of the world's energy, would be sufficient to head off global warming.[4]

This soak-the-rich strategy was seriously flawed from various points of view. Perhaps the most obvious is that we live in an era of global capitalism. Dirty industries, like all others, will relocate to places where they can be conducted most cheaply. Industries that were targets of emissions regulation would simply move to developing countries to reduce their costs and would continue to pump the same amount of carbon dioxide, carbon monoxide, and sulphur dioxide into the air.

Another complication is that the carbon dioxide level of the air is as much a function of plant life as it is of industry. In general, the more biomass on the earth, the more carbon dioxide gets used up. Destruction of rainforests, most of which are located in developing countries, subtracts huge amounts of living plants from the earth's biomass and thus increases carbon dioxide in the atmosphere. Developing nations are thus major contributors to the greenhouse effect even though they use relatively little energy. As envisaged, the Kyoto agreement could not work, even if members of the world community of nations agreed to participate in it.

Kyoto's biggest failing is that it ignores the lessons of the tragedy of the commons. The key to solving environmental problems is to demonstrate that restraint by individual countries ultimately contributes to their common good. If herdsmen could be assured that others would reduce their grazing, then the rational course would be to exercise restraint themselves to save the vital shared resource of the common pastures. That such efforts can work has been demonstrated repeatedly as a practical matter in fisheries management where fishermen voluntarily reduce their catches to avoid driving economically valued species to the verge of extinction.

Instead of emphasizing the shared interest of the world community in protecting the atmosphere, Kyoto began by driving a wedge between developed and developing nations. Why would some herdsmen voluntarily withdraw a sheep or a cow from the threatened pasture if others did not share the sacrifice? Doing so would be an indirect transfer of assets to the "cheats." The whole point of the tragedy of the commons is that the cost of restraint is borne by the individual, whereas the cost of cheating is borne by the community. The Kyoto agreement was set up so that some countries were given a license to cheat on the environment, whereas others were expected to bear the cost. The framers made the mistake of attempting to redistribute wealth from affluent countries to poor ones. This is a laudable goal but it is irrelevant to global warming. Even more to the point, wealthy countries do not want their "excess" wealth siphoned off to poorer nations and would never ratify an agreement that had this effect. If they did, they would consider themselves to be suckers!

Countries behave rather like individuals in international matters because important national decisions are made by political leaders, usually with an eye to the emotional reactions of their constituents. Individuals are capable of a very high level of altruism toward groups that they function in, from families to corporations and states, but their good deeds are contingent on evidence that others are behaving similarly. We never tolerate cheats gladly, and most cooperative endeavors are destroyed by defectors. Whether one considers the dynamics of life in small island communities or the defenses of industrialized countries against criminal enterprises, the response is always the same: identify the defectors and shut them out of the community. There has never been a community in which cooperators gladly coexisted with cheats.

Framers of the Kyoto agreement may have thought they would start something new. They were mistaken. Governments in affluent countries were unwilling to ratify the agreement. The US Senate rejected Kyoto by a unanimous vote because it failed to include restrictions on developing nations. An agreement that imposes restrictions on developed nations but allows poorer countries to experience all of the gain and none of the pain has just the kind of imbalance that our evolved mechanism of reciprocal altruism forces us to balk at. Even if it were palatable to economically developed countries, the Kyoto agreement, in its current form, would do little to curb carbon emissions because the most irresponsible destruction of the environment takes place in developing countries, where life is difficult and environmental concerns are of limited importance.

THE PRISONER'S DILEMMA OF GLOBAL POLLUTION

Global pollution is a tragedy of the commons. This seems depressing. Yet, tragedies of the commons can be solved. In reality very few people want the short-term benefits of cheating if they know that the long-term consequence will be ruinous for them as well as for other inhabitants of the planet.

This argument is not just theoretical. One of the most striking examples in recent history concerns the balance of nuclear terror between the United States and Russia. These superpowers trained their nuclear warheads on each other, each capable of taking out a large city and contaminating huge areas with nuclear dust that would make them practically uninhabitable for future generations.

In this prisoner's dilemma, both sides had the choice of initiating a nuclear strike (the defect strategy) or refraining from the first use of these devastating weapons (the cooperative strategy). Both sides knew that if they had launched an attack, it would have been swiftly followed by an all-out counterattack. Ordering a nuclear attack under such conditions was tantamount to committing suicide. Both sides knew that no first strike would be capable of destroying all of the enemy's strategic nuclear weapons and that launching an attack would be the height of folly. This philosophy, known as Mutual Assured Destruction, or MAD, guided the strategy of each side, preventing the use of nuclear weapons during the Cold War period.

It should be obvious that a similar logic applies to global pollution. The short-term immediate gains of national economies is balanced by a cumulative, and possibly irremedial, destruction of the global environment. The main logical difference, and it is clearly an important one, is that with a nuclear strike, Armageddon would have occurred within an hour. With the wanton destruction of the environment, mutual assured destruction is delayed, possibly for centuries. When the world community of nations is fully apprised of the facts—and this may take several decades as scientists produce reliable projections of the greenhouse effect—they will likely act in responsible ways to avert the threat.[5]

One of the fascinating aspects of the problem of global pollution is that nested within the global prisoner's dilemma, each nation has its own prisoner's dilemma in which the polluting activities of corporations and individuals degrade the quality of the air that they breathe in their home nation, as well as affecting the global atmosphere. The global citizenship of individual nations is very much determined by domestic conditions, particularly economic development. Poor countries are far more interested in developing dirty industries than in controlling pollution. In developed countries, people have better health, live longer, and are much more concerned about the quality of the environment as a health threat.

The split between global citizenship and national self-interest is clearly illustrated by the United States and its positions on the Kyoto agreement. The failed agreement called for restrictions that affected developed nations exclusively. The Kyoto Protocol called for nations of the industrialized world to reduce their carbon emissions to 5.2 percent below their 1990 level by 2012. As the largest energy user in the world, the United States agreed to reduce its greenhouse gases by 7 percent.[6]

Despite its resounding failure to ratify Kyoto, the United States continues to make strides in improving the quality of its air, particularly over congested urban districts, in compliance with increasingly stringent guidelines of the Clean Air Act of 1970 that is renewable every ten years. For example, sulfur dioxide emissions, the main cause of acid rain that damages plants and infrastructure, was reduced by one-third in the thirty years following passage of the act. What is more, legislation modeled after the Clean Air Act has been passed in many of the developed countries of the world, producing gradual improvements in the quality of urban air.[7]

The Clean Air Act was designed to reduce pollutants that undermine public health, rather than to address global warming, however. It controls six types of emissions: sulfur oxides, nitrogen oxides, hydrocarbons, carbon monoxide, photochemical oxidants, and particulates (i.e., dust), but not carbon dioxide. All of the controlled emissions contribute to city smog, which elevates mortality in cities and triggers asthma attacks. In some cities, like Los Angeles, children are encouraged to stay indoors on days when ozone levels in the air are high because developing lungs are particularly vulnerable to the adverse effects of air pollution. Smog levels are affected by weather conditions and are often worst during the summer months.[8]

Air pollution hits home, and that is why developed countries have taken firm steps to improve the quality of their air. From a national perspective, they are addressing the prisoner's dilemma of pollution by means of governmental regulation. Some of these regulations have been quite imaginative. Thus, the production of sulfur dioxide by the power generation industry—a major source of acid rain—was part of an innovative trading system according to which a company's ability to reduce sulfur dioxide emissions had a beneficial effect on its bottom line. In response, power producers shifted to a low-sulfur form of coal that produced a sharp decline in sulphur emissions.

Concern for the environment is much more pronounced in developed countries than it is in underdeveloped ones. This is true not just for domestic pollution, but also regarding damage to the global environment and ecosystems. This concern is demonstrated by the United States' initial willingness to go beyond the Kyoto guidelines, for example. Why have developed countries acted as more conscientious stewards of the environment than underdeveloped ones?

The short answer is that for many poor countries, and for the people in them, concern about pollution seems like a luxury they cannot afford. Environmental regulation increases the cost of doing business. Many dirty industries in the United States have consequently opted to relocate south of the border, in Mexico, rather than take on the additional expense of installing smokestack scrubbers to reduce emissions. The Mexican population was not offended by the arrival of so many polluting industries south of the Rio Grande: they welcomed the new enterprises as an infusion of much-needed jobs and capital investment.

In general, air pollution is not an inevitable consequence of heavy industry. It results, instead, from a lack of will, or money, to invest in technologies for reducing emissions. The will is weaker in developing countries because they have more pressing concerns. Their problems are immediate: creating employment, fighting disease epidemics like AIDS that is currently devastating the poor countries of Africa, and reacting to famines. In a country where the average person is lucky to make it to the age of forty years, worrying about the distant future when people might be sickened by pollution is far from being a priority.[9]

Concern with the environment is more typical of affluent countries, reflecting the development of their economies. Having gone from agricultural to industrial economies, and moved on from heavy industry to service industries, concern for the environment is a luxury they *can* afford. When England underwent its industrial revolution in the nineteenth century, for example, workers lived in the most appalling conditions, crowded together in cities that lacked sanitation, decimated by epidemic diseases, and suffering from a variety of ailments including repetitive stress disorders, injuries from machines, and breathing problems due to coal smoke. Dirty industry was accepted as an inevitable consequence of economic progress. The same attitude is prevalent in developing nations today. Even within the United States, some communities remain ambivalent about industrial pollution. Thus, in the comparatively poor state of Maine, many residents were remarkably tolerant of the highly polluting paper industry that fills communities with a horrifying odor similar to rotting eggs. Mainers referred to this as "the smell of money."

The most serious defect of the Kyoto agreement was that it failed to curb greenhouse gas emissions in developing countries. This omission seriously undermined its potential effectiveness even if it had been more palatable to wealthy countries. The problem is that developing nations of today are likely to be economic powerhouses of the future. Take China, the largest developing country at present. If current rates of economic growth continue, China is poised to become the world's largest consumer of energy by 2030.

This is a fairly frightening prospect to environmentalists for several reasons. Like many other developing nations, China uses energy inefficiently. When residents burn coal to keep themselves warm in the winter, most of the heat escapes through chimneys and poorly insulated walls and roofs. And China derives most of its energy from coal. Coal generates a lot of pollution when it is burned, emitting huge quantities of greenhouse gases. China's coal consumption is expected to double in the next twenty years, assuming it continues on its rapid economic growth trajectory. In the future, developing countries will be responsible for most of the world's air pollution because their populations are increasing rapidly and their economies are capable of rapid growth as they close economic gaps between themselves and the affluent countries of today through increased participation in global trade.[10]

By contrast, the contributions of developed nations to environmental pollution could well decline, in the absence of regulations, for various reasons. One is the decline in heavy industry, much of which is getting shifted to developing nations. Birth rates in many developed nations, particularly those of Europe, are currently well below replacement, which forecasts future economic decline, although serious drops in productive-age populations have been averted by immigration.[11]

Shrinking populations would inevitably result in declining energy consumption, all else being equal. Given that developed nations are currently the ones that are most concerned about environmental issues, the impetus of technological innovation for energy efficiency and pollution control will likely continue there. There are thus very good reasons for predicting that energy use, and atmospheric pollution, may decline in affluent countries but will rise in developing ones. If so, then the current omission of developing countries from the Kyoto restrictions is astonishingly short-sighted.

Even so, a reasonable case can be made for going easy on developing countries in respect to pollution control—provided that they restore endangered ecosystems to reduce atmospheric carbon. One example would be restricting logging and mining in the rainforests.

As countries become affluent they pass through a stage of industrialization, or heavy industry, that gives way to a service-based economy, in which most of the income is generated from activities like financial services, restaurants, medicine, social services, and government bureaucracy, rather than manufacturing. In the United States, for example, approximately 80 percent of economic production is in the service sector, with the remaining fifth accounting for manufacturing and agriculture.[12] Would it be fair to crimp the development of poor countries by imposing stiff environmental restrictions when the wealthy countries have already polluted their way to affluence? One reply to this argument is that the threat of global warming, for example, was unknown when England was in the heyday of its industrial revolution in the middle of the nineteenth century. The greenhouse effect is now a confirmed reality, according to the consensus of leading environmental scientists, a group that waffled over the issue for decades.[13] Now that we have been alerted to the problem, there is little excuse for doing nothing about it. Moreover, since nations who do nothing are defectors in the prisoner's dilemma of global pollution, and thus invite defection by others, any restriction of carbon emissions must apply to all countries if it is to be effective in changing the behavior of any.

The problem of crimping economic growth could also be negotiated because there are several ways of changing carbon levels in the atmosphere. Developing nations that find it difficult to regulate their developing industries for economic reasons could compensate by an aggressive promotion of fragile ecosystems, particularly rainforests, that are effective at sucking vast quantities of carbon dioxide out of the earth's atmosphere. Such flexibility is a hallmark of effective environmental regulation, judging from the success of the United States in reducing pollution from power producers with a market-based emissions trading system.[14]

Global environmental regulation is clearly desirable if we are to solve major ecological problems that threaten the stability and viability of our entire way of life. Yet, it may not happen, at least for a long time. This does not mean that the planet is doomed, because the tragedy of the commons of environmental pollution *within* nations is more easily solved through governmental regulations. If most of the governments of the world make their own urban atmospheres more livable, as seems likely, then the problem of global pollution could be controlled, although it may be much more difficult to deal with global warming because the immediate effects of pollution on local temperatures are so subtle. What is more, those populations most at risk, particularly low-lying island communities, which could be completely inundated in a couple of centuries, typically produce little in the way of greenhouse gases and are thus powerless to redress the problem.[15]

Prevention of global warming requires a deliberate switch to cleaner forms of energy, particularly hydroelectricity, solar energy, wind power, and other alternatives that produce no greenhouse gases. Alternatives to fossil fuels for power generation are generally much more expensive and have other drawbacks, such as unreliability or lack of capacity to satisfy high demand. Faced with this conflict between economic self-interest and damage to the global environment, most countries have followed the selfish route. In time, fossil fuels will run out and/or become more expensive to exploit, however. When that happens, a lot more effort will go into technological innovations for the efficient exploitation of alternative energy sources.

There are two kinds of commons problems in the degradation of the world's environment, the one that occurs within the community of nations, and the one that occurs within a country. Clearly, there is an important economic dimension that underlies the tragedy of global warming, in the sense that the self-interest of nations urges them to put more carbon dioxide into the global pasture, just like the herdsman being driven by economic self-interest to put out another sheep on the commons. Within countries, the tragedy of the commons of environmental pollution is also rooted in economic self-interest. Thus polluting industries profit while making people sick. Rather less obvious, but equally significant, is the fact that living the good life, as conceived in materialistic countries, pits the hedonism of the individual directly against the global environment.

THE GOOD LIFE VERSUS THE ENVIRONMENT

When we think of the Industrial Revolution in England, we think of foul-smelling Dickensian towns with smoke billowing from huge industrial chimneys and soot everywhere. Having largely passed on from heavy industry, modern economies are service-based, which suggests that they should be a lot less polluting. In reality, it is the wealthy countries that use most of the earth's energy. Thus, Americans use approximately six times as much electricity as the average person in the world, according to CIA data.[16] They also use far more oil to run their numerous vehicles. The affluent lifestyle is one that burns a lot of energy, thus contributing, almost inevitably, to environmental problems.

The fact that advanced economies have moved beyond the era of the industrial smokestack and that we live in service-based economies might seem to be good for the environment, but that is an illusion. Industrial pollution is conspicuous because it is highly concentrated at point sources, and may thus be damaging to the local ecology. Service economies do not use less energy than industrial ones; in fact, they use far more, as reflected in the US case.

Why do service economies use so much energy, thus contributing disproportionately to global warming? The simple answer is lifestyle. Take the financial services industry, for example. A financial company might seem like a clean industry but its office premises use a vast amount of energy for heat, light, and air conditioning. Since most of our energy is still derived from fossil fuel, these energy needs necessitate the burning of large amounts of oil, coal, and natural gas. Employees must also drive to work, often from distant suburban homes, further contributing to global warming from oil combustion, and to air pollution.

Financial industries, like all others, contribute to urban expansion. This means that trees and vegetation are replaced by tarmac and concrete. Paving is an important factor in local warming around cities. Thus, the average temperature in Japan increased by approximately one degree centigrade during the twentieth century, but the temperature in Tokyo increased by three degrees.[17] By the same token, high-technology industries, which bring in a substantial proportion of national income for small countries like Singapore and Ireland, as well as larger ones, like Japan and the United States, appear to be a great deal "greener" than they are in practice. Computers, particularly the large servers that run the Internet, consume large amounts of electricity. On the other hand, advances in electronic communication make it easier for people to work at a site that is distant from their home office, which allows for modest decentralization. If more people work from home, this reduces urban concentration of population and reduces commuter traffic. Enticing as this prospect is, most large cities have continued to grow relentlessly, except that the growth has shifted from the industrial centers to residential suburbs.

Progressive urbanization was the big theme of demographic

change throughout the twentieth century in industrializing countries. With mechanization of agriculture, many peasants lost their livelihood and were forced to relocate to cities in search of employment. Cities are places where large amounts of money get made and where standards of living increase from generation to generation, although this is not true of those who are desperately poor to begin with and thus ill equipped to exploit economic opportunities.

Whether workers feel better off or not, there has been a huge increase in real wages throughout the twentieth century in the United States and other wealthy nations. The average person has more disposable income, enjoys better health, and lives substantially longer. With urbanization, there has been a progressive increase in the number of affluent people, although the number living in wretched poverty has changed very little, due partly to a decline in demand for unskilled labor. With increased affluence comes the affluent lifestyle and increased consumption of energy, which is potentially damaging to the environment.

The affluent lifestyle is partly driven by social comparison. If my neighbor purchases a boat, then I might consider purchasing a larger boat to make it seem that I am more economically successful than he is. Yet, it cannot be denied that some people enjoy the activities that are possible for boat owners but denied to others, such as impromptu cocktail parties on the water, or visits to uninhabited islands. Even so, the main enjoyment of wealth could be ostentation. If you don't care what other people think of you, you can live quite happily in modest circumstances, if not, like Diogenes, in a barrel. This insight was articulated by early sociologist Thorstein Veblen (1857–1929), who coined the phrase "conspicuous consumption."[18] In the modern world, conspicuous consumption is less evident in the extravagant clothing of affluent women, which hindered their movements and made it impractical for them to do any useful work. Today, affluence is expressed instead in consumption of expensive goods and services, and thus in the consumption of energy.

Virtually all manufactured products involve the consumption of substantial amounts of energy, but the conspicuous consumption of energy is most easily illustrated in the case of large homes and vehicles. The affluent lifestyle involves movement in an astonishing variety of vehicles that are energy-expensive in terms of manufacture and operation. This is particularly obvious for Americans who own far more automobiles than is necessary for getting around. Many vehicles are designed and purchased primarily for leisure use, including golf carts, snow mobiles, jet skis, and most motor boats and sailing vessels. The prevalence of such recreational vehicles contributes to the huge domestic consumption of oil in the United States, which uses about a quarter of the world's oil. Two other prominent affluence-related trends contribute enormously to oil consumption. They are the increased size of modern homes and the popularity of gas-guzzling utility vehicles.

Owning a large home makes a statement about social status and is actually an important repository of wealth. In affluent societies, home size is likely to increase. Such increases can be substantial. In the United States, the average home size increased from approximately 1,500 square feet in 1970 to 2,000 square feet in 1994.[19] This is a remarkable change considering that family size declined during the same period. Large homes require large amounts of lumber and thus necessitate the felling of many trees, with implications for the greenhouse effect. All the other materials used in building consume considerable amounts of energy in their fabrication.

The energy cost associated with home construction is dwarfed by the energetic demand of heating and cooling, which increases as a function of home size. Even with improved thermal insulation of modern houses, the amount of oil required to heat a home during the winter may exceed one thousand gallons. Then there is the electricity required to run air conditioning systems in the summer. Given that 70 percent of US electricity is generated from fossil fuel,[20] electricity consumption inevitably means contributing to global warming.

Mindful of the ecological toll taken by modern housing, the "green" architecture movement has developed housing models that are minimally damaging to the environment in terms of materials used and energy conservation. Some of these homes use solar panels as a primary source of energy, which gets around the problem of having to burn fossil fuels. Others emphasize design features that minimize interior temperature fluctuation. Massive walls, sometimes constructed from old tires filled with soil, store heat during warm days and release it gradually at night, ensuring that temperature fluctuations are minimal. This cavelike approach to temperature regulation does not appeal to those who want their homes to be light and bright. Such esthetic proclivities mean that most people avoid using ecologically friendly housing style and put their happiness, or self-interest, before the good of the planet.[21]

The sport utility vehicle, or SUV, is another case where selfish design choices take precedence over the interests of the environment. Popular because of its spaciousness, power, and sense of security due to size, the SUV is bad for the environment because of the greater amount of energy required to haul a heavier vehicle that accelerates like a smaller car.

The success of larger vehicles in recent decades has been phenomenal. In Canada, for example, the number of SUVs, light trucks, and minivans exceeded cars in 1999. Sales of such heavier vehicles doubled in the final decade of the twentieth century. According to Canada's Office of Energy Efficiency, greenhouse gas emissions rose by 4.5 million tons each year due to the heavier vehicle fad. This preference more than offset efforts to improve the efficiency of other vehicles that amounted to a reduction of only 3.9 million tons a year. Heavier vehicles use about 70 percent more fuel per mile compared to small cars. These SUVs also pose a danger to smaller more fuel-efficient cars.[22]

The connection between affluence and environmental damage is inescapable. Does this mean that our economic greed is propelling us toward ecological collapse?

Countries where there is a high level of disposable income are characterized by service economies. By virtue of their affluent lifestyles, individuals in such societies expend huge amounts of energy, even compared to newly industrialized countries where factories belch thick smoke into the atmosphere. Thus, the 20 percent of the world's population living in affluent countries consumes 60 percent of the world's energy according to the United Nations

Department for Policy Coordination and Sustainable Development.[23] Americans are particularly high on energy use, even among developed countries. The average US resident uses twice as much electricity as the average French person, for example (approximately 13,000 KWH versus 6,600 KWH, based on data in the CIA's *World Factbook*).[24] What makes this statistic so impressive is that the average French person uses over three times as much electricity as the average person in the world.

Interestingly, affluent countries can use large amounts of energy without necessarily compromising the quality of their own air. This is well illustrated in the case of the United States, which has far better air quality than most industrial countries but nevertheless uses a substantial fraction of the world's energy. Wealthy countries deploy technologies of pollution control that poor countries cannot afford. They also use their energy more efficiently. For example, less fuel needs to be burned to heat a home because of superior insulation. With the emergence of hybrid cars, and fuel cell technology that is completely free of greenhouse gas emissions, the advantage of cleaner technologies becomes more significant. These technologies are quite expensive, however, so that the benefits will inevitably be enjoyed most by affluent nations.[25]

So far as the global tragedy of the commons is concerned, developed countries use most of the world's fuel and therefore contribute disproportionately to the greenhouse effect. Despite such damage to the global environment, affluent countries may not pollute the air or water within their national boundaries to the point that their national quality of life is undermined. It seems obvious that countries will continue to pursue their selfish interests to the detriment of the global environment, until a new and improved Kyoto-type agreement is devised that allows all nations to reduce greenhouse gas emissions, secure in the knowledge that their cooperative behavior is not being undermined by selfish defectors. That such agreements can work is illustrated by the success of global action against the hole in the ozone layer. Obviously, countries can cooperate when it is in their vital collective interest to do so.

Advanced service economies use far more energy than those in which most wealth derives from heavy industry. Whether energy use contributes to global warming is very much determined by the energy source, however. France is a typical service economy in many respects but it is distinguished by obtaining most of its electricity (76 percent) from nuclear power that does not increase the atmospheric load of carbon dioxide, although it incurs severe problems of radioactive pollution. Hydroelectricity is another key "alternative" energy source (i.e., one that avoids the equation between energy consumption and greenhouse gases). Very few developed countries exploit much hydroelectricity because of the lack of suitable rivers. One exception is Norway, which benefits from a physical geography suitable for harnessing hydroelectricity.

The affluent lifestyle is costly to the environment, and Veblen's concept of conspicuous consumption could as easily be applied to consumption of energy as money. The fact that, as individuals, we are such appalling stewards of the environment would seem like a conspicuous failing of the argument that people are biologically predisposed to altruism. On closer inspection, this is less of a problem because ancestral human populations rarely faced the kinds of ecological bind in which restraint of selfishness would have saved them from ecological collapse.

THE NOBLE SAVAGE SAVAGES THE ENVIRONMENT

The received wisdom within anthropology has long been that indigenous populations behaved in a manner that preserved their local ecology: they lived in balance with nature. Many chinks in this argument have emerged in the recent work of anthropologists and other scholars. We hear of hunters pursuing their game animals to the point of extinction, of horticulturalists destroying the rainforest, and of farmers turning their soil into toxic wastelands through salination caused by irrigation. Instead of conservation of precious game animal stocks, anthropologists have found signs of wanton butchery with little thought of the future.

The argument that early humans remained in balance with their prey animals does not mean that they were inspired by ideals of conservation. In the normal course of events for other predator species, there is a dynamic balance between the populations of predator and prey. If the number of predators rises quickly, they deplete the population of prey animals, which produces a crash in the number of predators due to the declining food supply.

Effective hunting is thus a curiously self-defeating exercise. If lions were to work hard at killing their prey, their food would tend to get scarcer the harder they worked. They avoid this problem by being exceptionally lazy, often lying in the shade for twenty hours a day. Human hunters also enjoy the leisure lifestyle, rarely working more than about five or six hours in a day.

For most of their history, the human species followed the rule that hunters should be lazy. Approximately forty thousand years ago, our ancestors improved their hunting technology and used bows and arrows and spear-launchers that allowed them to kill from a distance. Instead of scavenging on large animals that had died, or deliberately driving them over cliffs, our ancestors could launch attacks against large dangerous prey animals.

Why this technological change occurred is unknown, but some anthropologists believe that it reflected increasing social complexity, possibly based on language elaboration, and the emergence of food cultivation on a part-time basis.[26]

After our ancestors developed a better technology for slaughtering large wild animals, they went to town on their prey with all the abandon of US buffalo hunters of the 1850s; professional hunters like Buffalo Bill Cody satisfied the hunger of railroad workers—using highly accurate rifles—thus precipitating the utter collapse of a population of bison that had numbered about thirty million.

Signs of the same abandon are evident in our ancestors. As soon as their hunting technology improved, they evidently hunted most of their large game animals to extinction. The evidence for this claim is remarkably strong. Thus, about forty thousand years ago, soon after the arrival of humans in Australia, many large herbivores disappeared from the fossil record. Thirteen genera perished. They included a giant horned tortoise, a wombat as large as a rhinoceros, and the marsupial lion (*Thylacoleo*) that had preyed on them. When humans arrived in the Americas via the land bridge that then existed between Russia and Alaska, the same kind of ecological damage ensued. Such destructions of large animal populations is tellingly referred to as the Pleistocene overkill. Thirty-three out of forty-five genera of large animals (or 73 percent) disappeared. These extinctions, coinciding with the arrival of humans, can only be satisfactorily explained in terms of human activities.[27]

The capacity of humans to wipe out their prey animals may be due to improved hunting technology, but knowing this immediate cause cannot help us to understand the deeper question of how a predator can completely eradicate most of its prey animals without wiping itself out in the process.

English science writer and researcher Colin Tudge believes that the answer to this question is agriculture. He argues that cultivated food insulated our ancestors against the effects of their excessive hunting. Subsisting on the fruits of their labor, they could hunt their prey species to the verge of extinction without the usual checks and balances of predator-prey dynamics.

By killing off so many large animals, our environmentally insensitive ancestors may have committed their descendants to the sad horror of work on farms. Unlike the subsistence way of life, where laziness is adaptive, in agriculture, increased work effort produces a corresponding increase in food yield.

Hard-working farmers produce more food and can raise more children than lazy ones. The whole story of modern economic development is a series of such positive feedback loops that broke the ancestral equilibrium between human populations and their subsistence environment.

Increased agricultural food production allowed the human population to rise sharply, setting the stage for industrial development and global warming. Rising population created a demand for further increases in food production. Thus, advances in agricultural technology were accompanied by steady increases in the human population, setting the stage for the rise of modern cities and the Industrial Revolution.

The Pleistocene overkill suggests that our ancestors were shockingly indifferent to their effects on the environment. Not all human populations have been hostile to their local environments, or their prey animals, of course. Thus, many existing societies of hunter-gatherers practice elaborate religious rituals to avoid offending their prey animals. Some anthropologists have interpreted this prayerful attitude as evidence of a more global ecological consciousness. This could be a mistake. Ritual apologies to the prey species, and emphasis on ritual purity while actually engaged in the hunt, can be more convincingly explained as a mechanism for dealing with the uncertainty of the chase. Thus, many gamblers are superstitious. Their lucky rabbit's foot or their lucky numbers may have no real effect on the outcome of a game of chance, but these things do relieve the anxiety associated with uncertain endeavors.

In conclusion, there is little evidence of sound environmental stewardship among our human ancestors. There are two good reasons why concern for the environment has rarely been to the forefront in the human past. The first is that the hunter-gatherer way of life that sustained our ancestors for some two million years maintained a very low population density in which humans were no more of a threat to their local ecology than any other large predator. The second is that throughout our more recent history as a species—the past forty thousand years or so—technological innovations have spurred population increases in a feed-forward process that continues to the point of threatening the global ecology with irreversible damage, such as global warming. Throughout this process of economic, and social, development, ruthless exploitation of natural resources was rewarded at every step by increased food production and enhanced reproductive success. The global tragedy of the commons is thus a phenomenon that has surfaced only in the past century but has been in the works for millennia.

There are no human adaptations to save the environment from damage via human economic activity. There are, nevertheless, adaptations for altruism toward members of the local community. If we want to save the earth from ecological catastrophe, we must begin with the premise that all nations belong to a global village and that their membership in the world community obligates them to protect their common resources of atmosphere, oceans, and ecosystems. Such cooperation is necessary for communal survival. It succeeded in protecting the ozone layer, and it is obviously preferable to any alternative.

Bad environmental stewardship is better explained in terms of history than morality. Most of us do not want to destroy precious natural resources on which our own health and survival depend, as well as those of our families and friends. Such tragedies of the commons are an inevitable consequence of our history as a species. It is not, however, inevitable that we will allow the tragedy to proceed to its logical denouement.

Altruism fails in relation to the environment, just as altruism fails in relation to out-groups. In each case, these failures reflect limitations in the evolution of altruism, rather than the generalized moral failures to which they are often attributed by pessimists.

Sympathetically appraised, the evidence for an evolved propensity for altruism in human beings is compelling. Yet, there is a major stumbling block for some readers, even when they ignore the political pages of their newspapers and examine instead the heinous practices of certain individuals acting solely on their own behalf. If humans are designed by natural selection to behave in cooperative ways in their communities, how can there be so many vicious people around? The final chapter confronts the very ancient topic of human depravity, asking why some appalling people continue to flourish after millions of years of natural selection for cooperative behavior.

CHAPTER 12

Where Have All the Villains Gone?

Many religions recognize evil and struggle to explain its existence. In the Judeo-Christian tradition, sin is considered to be inborn. Original sin has much in common with the biological concept of cutthroat selfishness among competitors. When people behave selfishly, and are a threat to us personally, we often see them as evil. How can the existence of evil people be reconciled with adaptations for altruistic behavior?

EVIL IN EVOLUTIONARY PERSPECTIVE

Given that the biblical tradition has a great deal to say about original sin, it is remarkable that there is no clear definition of evil. Adam and Eve were banished from the good life in the Garden of Eden for eating from the Tree of Knowledge. What exactly this means remains murky. To many it implies excessive intellectual curiosity. The Shaking Quakers interpreted knowledge as carnal knowledge and jumped to the conclusion that human sexuality is the root of all evil. This interpretation is decidedly unorthodox. Christian theologians never condemned sexuality, as such, although many vehemently condemned sexual *pleasure*. Students of the Bible are unlikely to find much consistency in the behavioral definition of evil. Cain was very bad because he killed his brother, for example, but Abraham was considered saintly because of a willingness to execute his own son.

In the Old Testament, evil was implicitly defined as anything that went against the capricious wishes of the Israelites' ethnic deity. To this early definition of evil, many subtleties were later added, in the form of ethical principles of the Christian period, such as "love thine enemy." (Interestingly, many of the Christian ethical principles seem designed to combat the problems of out-group hostility, which might help to explain why Christianity has spread around the world so successfully.)

The Old Testament approach to evil, as disobedience, is a useful starting point. Modern secular definitions of evil focus on behavior that is rooted in the needs or whims of the individual and impervious to the requirements of the community. To some extent, evil is a matter of breaking laws laid down by the civil authorities. If one attempts to compile a list of exceptionally evil actions, most are serious breaches of the law. Consider the following categories of people that are commonly described as evil in journalistic descriptions:

- murderers
- serial killers
- serial rapists
- child molesters
- kidnappers
- terrorists
- muggers
- gangsters

Each of these categories of people is evil because each puts its selfish interests, such as money, the desire to kill, or sexual gratification, before the common good, which is impaired by the criminal actions. Undermining the common good is one defining feature of evil in a secular society. It may not be the most important criterion, however. We are more likely to see people as evil if they behave in ways that are severely threatening to ourselves, or to members of our families. Consider the following wrongdoers:

- insider traders
- computer hackers
- tax evaders
- corrupt accountants
- white-collar thieves

These crimes are common. They are devastating to national prosperity and involve far larger amounts of money than that which flows through the coffers of organized crime. Yet they are not generally seen as heinously evil, although this opinion may change as the number of people whose life savings were destroyed by the Enron debacle, and other such corporate frauds, becomes evident. White-collar criminals are perceived as less evil because they are considered less of a threat to us personally. Thus, they are rarely even prosecuted. White-collar crime is often difficult to detect, of course. When it *is* revealed, companies are sometimes happy to dismiss the wrongdoer, rather than invite the damaging publicity of criminal proceedings.

The actions of fraudulent company officers may also be hard to distinguish from those of executives who follow the law. Many of the supposed crimes of Enron executives, and others, revolve around a fine interpretation of accountancy rules that even corporate lawyers do not agree upon. When the fiber-optic telecommunications industry encountered rough times, for example, owners of cable networks inflated earnings by selling each other access to their networks. By this stratagem, both companies participating in the exchange sweetened their bottom line without producing any new goods or services. The same trick has been used by oil services companies involved in producing seismic maps that identify possible oil-drilling sites. These companies entered into partnerships to sell each other access to their seismic map collections. More blatant variations on this theme have been the sham sales of energy by Enron and others.

From one point of view, such transactions are quite silly. It is rather like a small child saying to his mother, "I made a lot of money this week because my friend gave me $10 if I would let him read all of my books. He is doing well too because I gave him $10 so I can read all of his stuff." Yet, the transactions were explicitly designed to defraud investors and succeeded in this respect even though no mother would be deceived by analogous claims from her child. Although such schemes were clearly intended to bamboozle, and defraud, investors, it is still not clear if they were actually illegal. Conceptually similar transactions that involve the sale of entire companies, that is, sham sales, apparently are illegal, which only contributes to the moral grayness of corporate law. Investors are not stupid, but it took the failure of large corporations to call their attention to the accounting shenanigans.

The moral ambiguity of white-collar crime is highlighted by the fact that in the very year preceding its collapse, Enron was universally recognized as a well-managed company. In a book published in 2001, John C. Maxwell, whose publisher claims that he is "America's expert on leadership," described Enron as "one of the best teams in the world."[1] Enron filed for bankruptcy protection in 2002, brought down by huge losses from investments in power companies and technology companies. The losses were concealed in off-balance-sheet partnerships, so that investors were deceived as to the true state of the company's books. Reported earnings were inflated by over a billion dollars between the third quarter of 2000 and the third quarter of 2001. Approximately 70 percent of the company's earnings were thus fictitious. The company was so crooked that even insiders longed for its demise.[2]

Far from being the ideal team that they seemed to Maxwell and other outsiders, Enron was spinning out of control. The innovative energy trading unit engaged in sham trades that artificially inflated company income and also conspired to manipulate power prices to its advantage. Of this moral and economic debacle, John C. Maxwell writes, "The president of Enron learned about the company's multimillion-dollar venture to go on-line only two months before the launch, and it didn't bother him a bit. Why? Because he and his team were reaping the benefits of the Law of the Compass" (i.e., that "vision gives team members direction and confidence"). When Maxwell picked Enron as the ideal example of his team-building principles, he indicated that it had its moral compass intact, as encapsulated in

Andrew Carnegie's claim that "A great business is seldom if ever built up, except on lines of strictest integrity." This is an odd claim coming from a robber baron who was one of the most inhumane employers in an era that treated laborers like draft animals.[3]

It is easy to make fun of Maxwell, who was spectacularly unfortunate in his choice of Enron as a poster company. Yet, it is clear that he was not alone in his admiration of the rapidly growing company. As Maxwell points out, Enron received numerous honors, including being number twenty-five on *Fortune* magazine's all-star list of global most admired companies (2000), and being named the world's leading power company by *Forbes Global Business* (1999). It was included twice in *Fortune*'s list of one hundred best companies to work for in America (1999, 2000), a bitter irony for those who lost their retirement savings that had been invested in company stock, as well as losing their jobs.[4]

After its financial collapse, the true nature of the Enron "team" came to light. According to William Powers Jr., Dean of the University of Texas Law School, who headed an investigation of the company's finances, there was a "systematic and pervasive attempt by Enron's management to misrepresent the company's financial condition."[5] Powers summed up the cause of the Enron fiasco as follows: "There was a fundamental default of leadership and management. Leadership and management begin at the top, with the CEO Kenneth Lay. In this company, leadership and management depended as well on the chief operating officer, Jeffrey Skilling. The board of directors failed in its duty to provide leadership and oversight."[6]

To paraphrase Dickens, it was the best of companies and the worst of companies. Evidently it is difficult to distinguish between aggressively growing, innovative companies and fraudulent enterprises.

Apart from such ambiguity, both moral and legal, another reason that white-collar crime flourishes is that its exponents are powerful people with connections in high places. Thus, the leading Enron executives had close ties to both President George W. Bush and Vice President Dick Cheney, both of whom had been executives in the energy industry in companies whose accounting practices were also questionable (i.e., Harkin Energy and Haliburton, respectively).[7]

Despite the grave economic consequences of corporate fraud, most of us see it as less evil than street crimes like mugging and bank robbery, that net only pennies by comparison, because street crime is threatening to us as individuals. It endangers life and limb. In addition to being personally threatening, evil acts are defined by selfish antisocial motives. People who are disordered, or insane, may not be seen as evil if their bad actions are out of their control.

MADNESS AND BADNESS

Modern biological psychiatry was founded on the discovery that insanity may be caused by brain disorders. The first compelling evidence for this emerged from the study of syphilis, a disease caused by bacterial infection. In the end stages of this disease, known as general paresis, the nervous system is attacked and psychosis, or severe loss of contact with reality, occurs. Typical symptoms include delusions of grandeur. When Richard von Krafft-Ebing demonstrated, in 1884, that general paresis patients were infected with syphilis, the connection between brain pathology and insanity was hard to dispute. Subsequent researchers showed that the syphilis bacterium, *Treponema pallidum*, is present in the brains of general paresis patients. Since this connection was made, there have been many efforts to show that bad behavior can result from brain pathology.[8]

Prior to the emergence of biological psychiatry, insanity was not clearly distinguishable from immoral behavior. Thus general paresis had been seen as a moral defect that was due to intemperate habits. Wilhelm Griesenger, a leading psychiatrist in Krafft-Ebing's day, felt that it was due mainly to inhaling smoke from cheap cigars, for example. The discoveries of modern biological psychiatry have suggested that some seemingly very evil actions are due to brain disorders.[9] If so, they cannot threaten the thesis that natural selection has designed people to behave in altruistic ways toward members of their communities.

Needless to say, the shift in description from bad to mad is highly controversial in many cases. Possible examples of this kind of ambiguous evil include the following:

- *Serial killers* who frequently act out irrational compulsions over which they seem to have diminished control. Some, like John Wayne Gacy, a sociopath, lead unexceptional lives apart from their serial killing, even to the extent of attending church and teaching Sunday school.
- *Mothers who kill their children* while suffering from postpartum psychosis. In some cases there is severe loss of contact with reality, to the point, for example, of believing that the children are possessed by demons. Such cases probably constitute a small minority of children killed by their mothers, however. Most mothers who intentionally kill their children do have severe psychological problems, however, particularly depression, anxiety, and substance abuse. Many are suicidal and kill their children because they do not want them to endure the pain of growing up without a mother. Many experience relationship problems, or are going through stressful divorces in which there is conflict over custody of the children. Some children are killed by their mothers to save them from physical, sexual, or emotional abuse. Physically abusive mothers may kill their children when attempts at discipline get out of hand due to poor impulse control on the part of the mother as well as lack of social support to relieve the stress of parenthood.[10]
- *Criminals with antisocial personalities* may inherit diminished sensitivity to their own immoral behavior. Formerly referred to as "moral imbeciles," they make up perhaps as much as half of the prison population. Sociopaths would likely have been at a severe disadvantage in the small communities in which our ancestors lived. They would have been ostracized, or even killed. It is thus reasonable to argue that antisocial personality is a genetic defect, conceptually similar to color blindness. Such defects can be maintained in populations at low levels if they are associated with recessive genes that are normally masked by the dominant allele. As usual, genetics is not the whole story. Moral intelligence may also fail to develop in individuals who have no genetic vulnerability but are raised by abusive adults.[11]

- *Rampage killers or mass murderers*. If antisocial personality is a brain disorder, this would appear to let half of the prison population off the moral hook, something that most people would probably be loath to do. Even more controversial is the view that high-school students who have gone on killing sprees aimed at classmates and teachers are psychiatrically ill. They are a particular case of rampage killing that occurs in other contexts, such as an employee killing coworkers.[12]

Many news accounts of high-school killers have exaggerated the ordinariness of their backgrounds. Thus, Kip Kinkel, an Oregonian who allegedly killed two classmates and wounded twenty-five, experienced auditory hallucinations, a defining characteristic of schizophrenia. One of his voices ordered him to kill. His family background was also replete with psychiatric disturbance, and five of his cousins had psychiatric diagnoses. In addition to these problems, and perhaps because of them, many high-school shooters were the victims of ridicule by peers, which provoked a misguided quest for respect.[13]

According to a *New York Times* analysis of one hundred rampage attacks, involving 102 killers, there was compelling evidence of psychiatric disturbance.[14] Almost half (47 percent) had a formal psychiatric diagnosis, a quarter (24 percent) were on medication, and 14 percent had stopped taking their medication immediately before the homicidal rampage. The fact that the majority did not have a psychiatric diagnosis is not firm evidence of normality: most of the people who are in need of psychiatric help may never receive it.

HOW CAN EVIL BEHAVIOR EVOLVE?

One of the most difficult conclusions of this book is that evil actions against out-groups are often motivated by altruistic tendencies toward members of an in-group. This argument sticks in the gullets of some readers but it is nevertheless strongly supported by much objective evidence. An evolutionary approach indicates that evil behavior can evolve if it is directed at competitive out-groups because such tendencies contribute to the survival, and reproductive success, of the kin groups and local communities on whose behalf they are exercised.

This thesis would appear to excuse all manner of atrocities committed in the name of warfare and political conflict. Yet, it does nothing of the sort. Ethnic conflicts of all kinds are tragic and, in most cases, tragically pointless. Instead of accepting the barbarity of war, civilized nations have come together to condemn atrocities and to set standards, based on the Nuremberg Code and the Geneva Convention, that protect the welfare of noncombatants and prisoners of war. Even though the medieval tradition of chivalry recognized some honorable principles in knightly contests, the previous history of warfare can be characterized as one of unrestricted savagery and butchery. In the case of the Nuremberg trials following World War II, and the recent trial of Slobodan Milosevic, the international community has shown a determination to punish political leaders for their crimes against humanity. Unfortunately, this sense of justice has been restricted to European nations and is blind to the ethnic cleansing that is an everyday feature of life in countries like Rwanda, in Africa, and Sri Lanka, in Asia. The Pol Pot massacre of political opponents in Cambodia was of a similar scale to Hitler's atrocities against the Jews but has not received as much attention. In any case, we can say that the glass of global political justice is half full, which is better than being empty.

Either way, the continuing existence of political atrocities cannot be used to falsify the argument that people have adaptations to promote altruistic behavior within their in-groups. That still leaves the considerable problem of evil action by criminals, and others, wherein individuals do harm to members of their own communities to promote their selfish interests. That such depravity exists can hardly be questioned. The real question is whether this fact falsifies evidence for the evolution of human altruism toward nonkin.

If evil is defined as destruction of the common good by acts of selfishness, three kinds of evil can be identified. There are the evil actions of the insane, the sane acts of violent criminals, and the calculated but physically nonthreatening acts such as white-collar crime.

Of these three categories, evil acts due to insanity present the least problem to an evolutionist. Mental illness, like other illnesses, is caused by the failure of an adaptive system. That psychotic women may murder their children, for example, does not in any way threaten the conclusion that there are evolved mechanisms motivating women to nurture their offspring. That such mechanisms may fail due to pathology does not deny that they exist. Evil due to madness is a failure of adaptive mechanisms that does not require additional evolutionary explanation.

This category may be expanded to include temporary insanity—the arousal of strong emotions, like sexual jealousy or fear. In some legal traditions, they are considered as having been committed under extenuating circumstances. Thus, in some societies, including Texas up to 1960, a man was permitted to kill his wife and her lover if he found them actually copulating. The rationale was that this provocation was unbearable and that the jealous rage it produced was uncontrollable.

Threatening violent acts by individuals who are not suffering from psychosis are more problematic. Violent criminals choose to defect, that is, to put personal gain before the common good and to destroy the quality of life of their victims without conscience or concern.

There are two basic reasons why individuals resort to a life of crime: genetics and environment. A substantial fraction of hardcore criminals have antisocial personalities, which may have a genetic component. Many of those with antisocial personalities operate in questionable businesses that are on the verge of fraud but technically legal, such as door-to-door sales of shoddy products at inflated prices, or on-line casinos that charge the stake but never pay the winnings. In this way, they can cheat the public without going to jail.[15]

Many criminals also commit crime because they are raised in environments where such "defection" is common. They are socialized for tough-mindedness, and selfishness, because those qualities are at a premium for survival under stressful social conditions, such as life in a slum. Needless to say, there is great variation in the level of altruism, and selfishness, that is inculcated in different societies, and in different households in the same society. Even in slums, people in one home may fight all the time, and in another they may get along fine. That helps explain why criminal behavior is found at all income levels.

The fact that some early environments reduce altruistic tendencies hardly threatens the assertion that in most environments, particularly those that approximate our hunter-gatherer origins, a high level of altruism is normal. With economic development, communal ties are weakened, and children learn to be more selfish, perhaps because this mindset helps them to succeed in a highly competitive economic environment.

Even though the early environment can promote criminal deviance, it is important to realize that criminals are not devoid of altruistic tendencies. Some may be generous to a fault. They treat their victims as an out-group, while preserving altruistic relationships with intimates. Bank robber John Dillinger was unusual in attempting to safeguard bank employees during hold-ups. By thus focusing his attacks on the hated institutions of the banks, he turned himself into a folk hero. He thus squared the circle by turning himself into a violent criminal who was loved, rather than feared.

John Dillinger's story shows that a violent criminal may direct altruism toward friends and relatives. Violent criminals, *as a group*, are less altruistic toward family members, however, and they have greatly diminished empathy toward strangers, particularly those that fall into the category of their victims. (We know that criminals are less altruistic toward family members because many are diagnosable as antisocial personalities, one of the defining characteristics of whom is cruelty and neglect toward children.)

As a group, violent criminals suffer from mental illnesses, or they had abusive backgrounds, or both. Yet, some who are neither severely mentally ill nor depressed deliberately engage in criminal behavior when they can get away with it.

In crimes of violence, the criminal elevates his, or her, selfish interests over those of the victim and the community. Such actions are exceptions that prove the rule of altruism on the part of law-abiding citizens. They are very damaging but they are also perpetrated by a small minority of individuals.

The fact that crime increases opportunistically during times of civil unrest is sometimes used as ammunition by those who would maintain, with Thomas Hobbes, that human beings are naturally depraved and that their evil character is revealed when the rule of law breaks down. This argument is a fallacy because human societies can exist, and have existed, witness Pitcairn Island, without any police force, or any external authority to bring people into line with the requirements of their community. For such natural social controls to be effective, people need to know each other well and have an expectation of interacting with each other for a long time into the future, however. Such conditions do not apply in modern societies where strangers constantly meet, thus necessitating the rule of law to restrain antisocial impulses that are more likely because they are directed at strangers.

Crime increases during times of war because civil authority declines, thereby increasing vulnerability to anonymous criminals. It is not simply a question of lack of protection by the civil authorities from unruly elements: criminals are themselves inhibited from committing crime if they think they will be recognized. Conversely, if people think that their individual actions are concealed when they participate in an unruly mob, their behavior becomes more antisocial. This phenomenon is illustrated by soccer hooliganism and the outbreaks of spontaneous looting that occur during riots, or even during a power outage. Nature is red in tooth and claw unless it is restrained by adaptations for altruism. Such altruism is most likely to occur among individuals who are well known to each other and is most likely to fail during interactions with strangers.

Violent crime is one challenge to the thesis of an evolved human altruism that extends beyond families. Crime represents the intrusion of biological selfishness where there is insufficient protection from altruistic adaptations, for example, because the victim is unknown to the perpetrator who cannot thus be identified and held responsible. (In some cases, however, criminals are known to victims of crime, and this is true for many crimes of passion that are provoked by strong emotions rather than a calculation of personal gain.) Violent criminals compose a small minority, many of whom could be categorized as moral imbeciles, lacking the basic psychological equipment for participation in normal altruistic interactions. A different sort of challenge is posed by more prevalent, but less violent, kinds of defection, such as white-collar crime.

WHITE-COLLAR CRIME

Asking why seemingly decent people commit nonviolent crimes is rather like asking why the chicken crossed the road. White-collar criminals are motivated primarily by greed, by the ill-gotten gains. They want to cross the road. On the other side are larger homes, faster cars, more desirable dates, and all the attractive illusions of opulence and comfort.

People obviously commit white-collar crimes for the economic payoff. The underlying evolutionary cause is competition over resources that limit reproductive success (specifically wealth). Economic competition is roughly equivalent to the selfishness that biologists see as the original state of nature out of which the various types of altruism arose.

Economic competition is a fact of life. Economists, in the tradition of Adam Smith, see it as a creative force that drives efficiency. It also has undesirable consequences, including the persistence of extreme poverty in affluent countries. Whether it is desirable or not, economic competition is a deeply irrational tendency that persists regardless of one's current level of wealth.

In the past, men were more acquisitive than women because economic success was an entry fee into marriage and reproduction, which were essential ingredients of social success and respectability. As described by Laura Betzig of the University of Michigan, wealthy men in many societies have been successful polygamists.[16] At an extreme were monarchs, like Solomon or Ismail the Bloodthirsty, who maintained harems of many hundred wives. Extraordinary wealth was thus converted into remarkable reproductive success.

Even though access to numerous sexually receptive women is a fantasy of most red-blooded men, its realization does not contribute to happiness. Thus, the Chinese emperors had a well-organized bureaucracy for servicing their wives according to a schedule that increased their probability of becoming pregnant. In effect, the emperor was thus turned into the stud animal on a baby-producing farm. Some found the schedule wearying and complained about the burden of their matrimonial duties.

Getting what you want is evidently not the key to happiness. By the same token, economic achievement does not satisfy the desire for wealth but may have the opposite effect of increasing greed. As individuals acquire greater wealth, they associate with richer people, thereby potentially exacerbating feelings of economic inferiority. These irrational aspects of greed help us to understand some of the bizarre excesses of white-collar criminals in recent history.

White-collar criminals are challenging to any thesis of universal human altruism, not because they look so evil, but because they look so ordinary. If people who dress in suits and get up early in the morning to attend their offices can be so dishonest, how can anyone be trusted?

A short answer is that kindness exists in a cruel world. Human altruism is tempered not just by competition between groups, but by competition between individuals. This is a difficult concept but it describes our behavior in the real world. When we go shopping, for example, our competitive knives are honed. We may refuse to buy gas at one station because it is seven cents dearer per gallon than another. While pumping gas at the cheaper station, we might blow our entire savings by contributing a dollar to a charity. Competitiveness in one sphere does not rule out its opposite in the other. In fact, the very cheapness of the individual in his business dealings may make him more likely to contribute to charity. Thus Bill Gates, one of the most voracious capitalists of our time, also contributes the largest sums to charity by funding his foundation. He is not unusual in this split between competitiveness and generosity. The same was true of the nineteenth-century robber barons, such as Dale Carnegie, who enslaved thousands in his steel mills, only to turn the ill-gotten gains into public libraries where the downtrodden could improve their minds.

Business ethicists have long recognized that their field poses unusual moral challenges. In most other occupations, people can maintain a code of personal ethics, often based on religion or ethical philosophies, that prohibits lying and cheating. In a 1968 article in the *Harvard Business Review*, Alfred Z. Carr concluded that ". . . the ethics of business are game ethics, different from the ethics of religion."[17]

According to personal (or religious) ethics, it is wrong to deceive others, wrong to steal, and wrong to exploit another person's failings and weaknesses. In game ethics, the rules of the game must be followed, but competitive players are not only tough on their opponents, looking for weaknesses and deliberately exploiting them, but they also know how to bend the rules to their advantage. A skilled competitive tennis player deceives his opponent into thinking a shot will be long when it is short. He deliberately plays into an opponent's backhand, if it is weak. If the opponent is not physically fit, he forces him to run all over the court to wear him out. If he is under pressure, and out of breath, he may challenge a line call merely for the opportunity to catch his breath. None of these actions is consistent with personal ethics but all are considered acceptable, or even admirable, aspects of gamesmanship. People who succeed at the highest level in sporting competition must be good at gamesmanship and even capable of bending the rules to their advantage.

The point of the analogy between business and games is that business is a competition. According to Quinn McKay, a business ethicist at the University of Utah, "If you're going to compete and win [in business], I think you have to practice gaming ethics."[18] The moral grayness of business ethics is illustrated by the deceptive practices of many successful businesspeople. John D. Rockefeller conspired with the railroads to cripple his competitors under the aegis of the South Improvement Company. In 1872, he lied under oath about his connection to the company.[19]

Rising entrepreneurs are also helped by lies and half-truths. In 1987, Kathy Taggares was working for Chef Ready Foods, a maker of frozen foods. She couldn't wait to leave and found her own business. Taggares privately negotiated with Marriott International to purchase one of its salad dressing factories. Her overture was warmly received but Marriott thought she was acting for Chef rather than herself. She did not disabuse them. By the time they found out, the deal had almost gone through and they accepted it. Taggares's business flourished and had over 350 employees by 1999.[20]

Actual lying is against the rules of business and may be illegal as well. Distortion and exaggeration is another matter. Most aggressive sales operations have elements of dishonesty, as most people recognize. The good qualities of the product being puffed are exaggerated and the weaknesses are ignored.

Lying, and theater, may be essential aspects of getting a new business off the ground. Someone who is starting a business from her home apartment may disguise the fact by using a more prestigious business address to receive mail. A one-person operator may conceal her lack of employees by always referring to her activities in the plural. "We will have your order shipped by next week." Skilled players of this game may use recorded background noise from a busy office to reassure clients over the telephone that this is a thriving business.

Such theater is particularly common in new businesses struggling to gain credibility. Without "creativity," it may be impossible for new entrants to bridge the credibility gap and convince clients to do business with them. The gap between the present state of a business and its future potential may be creatively filled for the benefit of potential investors as well as potential clients. Such "vision" is a feature of most successful entrepreneurs. To the extent that they are laying out their expectations for the future growth of a business, it may be sincere. Nevertheless, predicting the future is a lot like writing fiction: there is a great deal of leeway for deliberate exaggeration.

It is a truism of business history that when the bold schemes of entrepreneurs succeed, their distortions of the truth are not held against them. Rather, they are admired, in the same way that fans appreciate a clever feint on the football field that makes defenders collide with each other while the opponent sneaks around them and scores. When a business collapses, hurting many employees, investors, and creditors, it is another story.

It may seem charming for a solitary person running a business to play busy office sounds for the benefit of callers. Such ruses are far less appealing when they are employed by large companies. When

Wall Street analysts went on site visits to Enron's headquarters in Houston, they witnessed a dizzying spectacle of huge energy trades. The trouble was that employees were making fake calls to negotiate fictitious deals. It was all smoke and mirrors.[21]

Steeped in such an environment of institutional deception, it is perhaps understandable that the executives who presided over this gigantic fraud should believe that, as individuals, they had done nothing reprehensible. Does the moral fog of gaming ethics prevent business leaders from recognizing when they have strayed into clearly unethical activities? Can they detect when they have penetrated the fuzzy boundary of business ethics and crossed over to frank illegality?

The temptation to stretch the truth as a way of manipulating public confidence is acute when companies are first launched, and also after they have begun to fail. Just as young businesses may exaggerate their future prospects, so failing ones exaggerate how well they are doing at present. We are inclined to think of such accounting dishonesty as a recent invention, but it is not. In 1937, for example, a drug company, McKesson and Robbins, reported revenues of $10 million from a fictitious division.[22] At Enron, a slightly more creative approach was adopted for cooking the books. Numerous off-balance-sheet "partnerships" were created that could make huge losses without detracting from the impression that Enron was in excellent financial health.[23]

It is interesting that Jeffrey Skilling, the chief architect of the partnerships, evidently believes that he has done nothing wrong. This conviction allowed him to give testimony before Congress, however lacking in substance, when most of the other principals took the fifth. Some commentators see this attitude as part of the arrogance that flourished in Enron, along with the belief that they were devising new business models that allowed them to make up the rules as they went along. Such arrogance was fueled by personal economic success. Skilling is estimated to have netted $90 million from his years at Enron while his trusted subordinate Andrew Fastow, the chief financial officer who set up the off-balance-sheet partnerships, allegedly reaped some $40 million.[24]

Among its other marks of distinction, Enron is a spectacular case of the moral relativity of success and failure in business. Now that it has failed, the company is seen as having been inherently corrupt. Just a year before its bankruptcy, Enron was held up as a shining example of American enterprise.

If it is so hard to distinguish a really corrupt company from one that sets a shining example of business ethics, then the layperson can be excused for drawing the conclusion that ethics is not, or was not, a very important determinant of the behavior of executives in top businesses. If that is the case, then there is little that distinguishes the behavior of ordinary executives from those who are subsequently identified as white-collar criminals because their business happens to fail, bringing ruin to tens of thousands of people. We can only hope that business ethicists of the future are more successful in drawing lines between acceptable and immoral behavior. The current metaphor of game ethics is not reassuring, suggesting as it does that winning is more important than what one does along the way to victory.

The fundamental paradox of business ethics is that companies must compete with rivals, a situation that evokes out-group hostility. To the extent that one company may drive another out of business, this is more like warfare than sporting competition, where conflicts are more symbolic. When executives are placed in this competitive arena, it may not be realistic to expect that they would have much of an opportunity to express the evolved altruism that is normally evoked when people interact in small, friendly groups whose members have stable, trusting relationships with each other. The environment of modern corporations may be an imperfect analogy to life in a hunter-gather community in this respect. If so, how important is the evolutionary history of human altruism for modern economic realities?

EVOLVED ALTRUISM IN THE MODERN WORLD

One of the most shocking aspects of modern economic development is that there is an inverse relationship between the wealth of a country and the altruism of its children, as already mentioned. Children are essentially trained to be selfish in affluent societies because much is given to them and very little is asked in return. Altruism is very much a matter of learning to respond to social expectations: if nothing is asked of children, they cannot develop in an altruistic direction. That is why children who work long hours on their family farms have a much more altruistic disposition than urban children growing up in the lap of luxury in developed countries.

It is ironic that we see it as a shame when children are not given opportunities to develop their intellectual talents, as well as musical and artistic abilities, and sports skills, but few complain when children are not given responsibilities, like domestic chores, that allow them to contribute in meaningful ways to their households, and develop altruistic attitudes. In many cases, children are not even expected to be respectful toward their parents. Such pampering creates needy adults who ask for a lot but give little in return. This would be a recipe for disaster in a traditional society but is more functional in a market economy where individuals must be self-assertive, and self-promotional, to succeed in businesses and careers.

Most parents pamper their firstborn more than later children for the simple reason that additional children create a drain on parental attention. It is interesting that firstborns are more competitive and more success-oriented. Students of birth order have found that prominent men of history are far more likely to have been firstborn, reflecting advantages in parental investment, as well, perhaps, as experience in manipulating younger siblings.[25]

Only children are particularly likely to be overindulged and thereby to develop a high opinion of themselves. This is true even in developing countries, like China, where there is a strong tradition of reverence toward parents. With the one-child policy, which is more rigidly enforced in urban than rural areas, parents invest all their affection in a single descendant, more often a boy, who is thus horribly spoiled, to the horror of elderly observers who grew up in large families and had much expected of them. This illustrates the complex connection between the rearing environment and a child's altruism.

Human altruism is thus quite sensitive to the social environment. It survives in developed economies, of course, but it is weakened by the lack of close relationships between people and their neighbors. This is important because throughout our evolutionary history we have behaved most altruistically toward those with whom we have lived in close proximity. That is why many urbanites express a hankering for community but cannot establish stable social relationships, or even minimal interactions, with the horde of strangers by whom they are surrounded. Their communities are established instead through friendship networks that are spatially dispersed.

One distinctive aspect of life in hunter-gatherer communities is that a small group of people, numbering from about forty to one hundred individuals, spend much of their time in close proximity and develop strong personal ties of friendship and cooperation. This means that we function most effectively in comparatively small groups, even in modern societies. With the transition to agriculture, groups became much larger. African agricultural communities, consisting of clusters of villages, may number several thousand people, all of whom are known to each other.

Even with the transition to larger agricultural communities, altruistic relations among neighbors were preserved. As agricultural communities switch to a monetary economy, however, local altruism declines. Both points are illustrated by the Igbo tribe of southeastern Nigeria. The Igbo, who are united by a common language, are ten million strong. They subsist primarily through farming and fishing, but also do a little trading, metal work, shoemending, and tailoring. Population density is very high, up to one thousand people per square mile and village communities contain about two thousand individuals.[26]

Traditional Igbo society is built on communal principles in which individuals have many responsibilities to the community but can also expect to be helped in their hour of need. Igbo communalism is well expressed in their system of property ownership. All of the land is owned by the entire community, but, within the community, adults own land and enjoy security of tenure on their plots.

Igbo society is regulated by a system of reciprocity that is in many ways the opposite of the money-regulated society that we have come to see as normal. The Igbo way of doing things presupposes that individuals are basically altruistic and will want to help others when given an opportunity to do so. This is the antithesis of the assumption, by neoclassical economists, that people are intrinsically selfish and will act only to improve their personal well-being. It is interesting that with the development of a monetary economy, Igbo altruism has declined and they have started to behave with the selfishness expected by economists.

The Igbo not only operated largely outside the monetary economy, but they despised the implications of working for money. Up to about 1970, almost all labor was unpaid. During busy times, farmers simply asked for help from their kin who worked on a purely voluntary basis. It was considered very bad form to accept money in return for helping a fellow villager because monetary transactions tend to undermine the whole system of altruistic exchanges on which the Igbo economy, and society, were based.

Community as an alternative to economic transactions was evident in all of the Igbo's most important transactions. In developed countries, a couple's home is their most valuable possession, and they often labor for twenty or thirty years to buy it out. When villagers built a new house, the labor was supplied free by relatives and friends using free locally-available materials like sticks and mud. The only cost of their new home was the meal that they provided for builders at the end of the day's labors.

In developed countries, once people have acquired property, they generally purchase insurance to protect it from catastrophic loss. The Igbo did not need financial insurance because they knew that when disaster struck, their fellow villagers would be there to help. If a home was destroyed by fire or flood, for example, villagers would promptly rebuild it. When a farmer's crop produced a poor yield, others helped out by donating food. This is in stark contrast to the Igbo of today: when crops fail, Igbo farmers now starve.

Igbo society used to be administratively self-contained, in the sense that disputes and grievances were resolved with the help of tribal elders. Litigation in the courts was avoided. Taking another villager to court was considered both ridiculous and an act of treachery against the community. Those who called the police on their neighbors were sanctioned by the village council, and could be excluded from community celebrations, or forced to pay fines. In many cases, the community pressure was strong enough to force a villager to get charges against another dropped.

Life in an Igbo community clearly supported a high degree of altruism within the village. Much of this giving was spontaneous, or voluntary, because there were no sanctions against those who did not help in a particular situation. Instead, there was an understanding that the individual derives his, or her, strength from participation in the community. Those who helped out a lot were given status as elders and were recipients of admiration and respect but obtained no material rewards for their altruism.[27]

The decline in Igbo altruism corresponded historically with the introduction of wage labor on farms. This marked the breakdown of labor reciprocity. Villagers lost interest in providing free labor to their relatives. Farmers who needed help during busy seasons were forced to hire outside help.

With the collapse of the communal labor pool, other aspects of Igbo community fell into decay. Victims of fire, flood, and famine found that they no longer had a community support system and were left to cope as best they could.

Along with the decline in group reciprocity in labor exchanges and relief from emergencies, there was a rise in litigation. Villagers increasingly resorted to the police and the court system for mediating disputes. The fundamental reason for switching to the external legal system is that during recent history, elders have been selected on the basis of their personal wealth, rather than service to the community. Many obtained their riches by foul means and have been fond of taking bribes, so that their impartiality can rarely be trusted.

Incomplete though it is, the Igbo story speaks volumes about the fragility of altruistic social systems that may exist for thousands of years and disintegrate in a single generation. Arguing that humans have adaptations for group altruism is not a naively optimistic perspective. Adaptations are defined as much by the environment as they are by the organism. An adaptation means, that an animal, plant, or microbe, is suited to the way of life in its current habitat.

Changes in the habitat may make the adaptation useless. If giraffes were transported to treeless prairies, their long necks would be a useless inconvenience. So it is with altruism. Having a capacity to engage in reciprocal altruism is useful only if it occurs in a social environment that is full of cooperators.

In the Igbo case, it seems that their transition to a monetary economy undermined the spirit of cooperation that had held the community together. Monetary transactions open up communities to outside interactions, however. If farmers cannot rely on free labor from their kin, they must hire laborers from outside their community. This introduces an element of competition because outside laborers may avoid working hard and may do work of inferior quality, thus producing disagreement as to how much wages the farmer should pay.

Even though altruism is fragile, it is also remarkably persistent. The reciprocity that sustains hunter-gatherer communities survived the transition to agriculture, as the life style of the traditional Igbo movingly illustrates. Many of the functions of an Igbo community have re-emerged in state-level urban societies as well, with an increased sense of responsibility for the unfortunate being expressed through government policies. We live in an age of great optimism in the sense that seemingly intractable problems of the past are being addressed through the concerted efforts of citizens as expressed through their governments. In developed countries, education has become a universal right. In many of the democracies of Europe, universal health care is also a right so that ordinary people do not have to fear bankruptcy as a consequence of the misfortune of being seriously ill. Even the perennial problem of poverty can be addressed through the altruistic efforts of the electorate as expressed through their governments: among the social democracies of Europe, for example, single mothers of young children are shielded from economic deprivation perhaps for the first time in human history. Even in the United States, where government aid to the poor generally lags other developed countries, senior citizens enjoy an unprecedented level of economic well-being thanks to social security payments based on worker contributions.

Whether through payment of taxes, or volunteer service, most of us contribute to the common good in practically important ways. It is true that our society contains a minority of freeloaders, cheats, and villains who detract from the common good. In some cases, particularly white-collar crime, cheats are very difficult to detect and thus operate with near-impunity for decades. They are like the common cold of modern societies, an inconvenience that saps energy but is not life-threatening to our altruistic objectives.

A great deal of evidence has been presented showing that people living in monetary economies are capable of a high level of voluntary altruistic behavior, whether they are donating to charities, giving blood, volunteering time, or simply giving the time of day to strangers they encounter. It is also true that many aspects of modern life promote social isolation and destroy the fabric of reciprocity.

Despite the fragility of altruistic systems in particular cases, the desire to contribute to others is a powerful one, and an important ingredient of human happiness, and health. Medical research finds that social interaction, including physical contact, from the cradle to the grave is a critical ingredient of good immune function, health, and longevity. Incubator babies who are handled often grow more rapidly. Married people enjoy much better health. Those who are deeply involved with their local communities live longer lives.

Wherever we look, there is compelling evidence for human altruism as an evolved disposition. It expresses itself in a cruel world where the forces of competition can destroy it at any time. Yet, we can be as sure of its permanence as we are sure that the wild flowers will bloom next spring.

CHAPTER 1: ALTRUISM

1. T. R. Malthus, *An Essay on the Principle of Population* (London: Johnson, 1798).

2. "Scott, 'Beat Darwin by 65 Years,'" *Press Association*, October 15, 2003.

3. Charles Darwin, *On the Origin of Species by Means of Natural Selection* (London: Murray, 1859).

4. William D. Hamilton, "The Evolution of Social Behavior," *Journal of Theoretical Biology* 7 (1964): 1–52.

5. T. A. Evans, E. J. Wallis, and M. A. Elgar, "Making a Meal of Mother," *Nature* 376 (July 27, 1995): 299.

6. J. Alcock, *Animal Behavior: An Evolutionary Approach*, 4th ed. (Sunderland, MA: Sinauer, 1989); *The Triumph of Sociobiology* (New York: Oxford University Press, 2001).

7. Ibid.

8. R. L. Trivers, "Parent-Offspring Conflict," *American Zoologist* 14 (1974): 249–64.

9. N. Barber, *Encyclopedia of Ethics in Science and Technology* (New York: Facts on File, 2002).

10. P. A. Gowaty, "Daughters Dearest," *Natural History* 97, no. 4 (1988): 80–81; L. Cronk, *That Complex Whole: Culture and the Evolution of Human Behavior* (Boulder, CO: Westview, 1999).

11. Ibid.

12. K. Ivey, "Cooperative Reproduction in Ituri Forest Hunter-Gatherers: Who Cares for Efe Infants?" *Current Anthropology* 41 (2000): 856–65.

13. Hamilton, "The Evolution of Social Behavior," pp. 1–52.

14. R. L. Trivers and H. Hare, "Haplodiploidy and the Evolution of Social Insects," *Science* 191 (1976): 249–63.

15. P. W. Sherman, J. U. M. Jarvis, and S. H. Braude, "Naked Mole Rats," *Scientific American* 267, no. 2 (1992): 72–78.

16. K. Hawkes, "Kin Selection and Culture," *American Ethnologist* 10 (1983): 345–63.

17. R. B. Hames, "Relatedness and Interaction among the Ye'Kwana: A Preliminary Analysis," in *Evolutionary Biology and Human Social Behavior*, ed. N. A. Chagnon and W. Irons (North Scituate, MA: Duxbury, 1979), pp. 238–49.

18. N. A. Chagnon, "Mate Competition Favoring Close Kin and Village Fissioning among the Yanomamo Indians," in Chagnon and Irons, *Evolutionary Biology and Human Social Behavior*, pp. 86–132.

19. A. Jolly, *The Evolution of Primate Behavior*, 2d ed. (New York: Macmillan, 1985).

20. D. S. Judge and S. B. Hrdy, "Allocation of Accumulated Resources among Close Kin: Inheritance in Sacramento, California, 1890–1984," *Ethology and Sociobiology* 13 (1992): 495–522; M. S. Smith, B. J. Kish, and C. B. Crawford, "Inheritance of Wealth as Human Kin Investment," *Ethology and Sociobiology* 8 (1987): 171–82.

21. D. F. Bjorklund and A. D. Pellegrini, *The Origins of Human Nature: Evolutionary Developmental Psychology* (Washington, DC: American Psychological Association, 2002).

22. K. G. Anderson, H. Kaplan, and J. Lancaster, "Paternal Care by Genetic Fathers and Stepfathers I: Reports from Albuquerque Men," *Evolution and Human Behavior* 20 (1999): 405–31; K. G. Anderson et al., "Paternal Care by Genetic Fathers and Stepfathers II: Reports by Xhosa High-School Students," *Evolution and Human Behavior* 20 (1999): 432–51.

23. F. Marlowe, "Showoffs or Providers? The Parenting Effort of Hadza Men," *Evolution and Human Behavior* 20 (1999): 391–401.

24. N. Barber, "Machiavellianism and Altruism: Effect of Relatedness of Target Person on Machiavellian and Helping Attitudes," *Psychological Reports* 75 (1994): 403–22.

25. W. Jankowiak and M. Diderich, "Sibling Solidarity in a Polygamous Community in the USA: Unpacking Inclusive Fitness," *Evolution and Human Behavior* 21 (2000): 128–39.

26. Bjorklund and Pellegrini, *The Origins of Human Nature*.

27. N. Barber, *The Science of Romance* (Amherst, NY: Prometheus Books, 2002).

28. Judge and Hrdy, "Allocation of Accumulated Resources among Close Kin," pp. 495–522; Smith, Kish, and Crawford, "Inheritance of Wealth as Human Kin Investment," pp. 171–82.

CHAPTER 2: EVOLUTION AND ETHICS

1. W. D. Hamilton, "The Genetical Evolution of Social Behaviour," *Journal of Theoretical Biology* 7 (1964): 1–52.
2. P. W. Sherman, "Nepotism and the Evolution of Alarm Calls," *Science* 197 (1977): 1247–53.
3. D. W. Pfennig, "Kinship and Cannibalism," *BioScience* 47 (1997): 667–75.
4. Ibid.
5. K. Ivey, "Cooperative Reproduction in Ituri Forest Hunter-Gatherers: Who Cares for Efe Infants?" *Current Anthropology* 41 (2000): 856–65.
6. M. W. Wiederman, "Extramarital Sex: Prevalence and Correlates in a National Survey," *Journal of Sex Research* 34 (1997): 167–74.
7. R. D. Alexander, *Darwinism and Human Affairs* (Seattle: University of Washington Press, 1979).
8. R. L. Trivers, "The Evolution of Reciprocal Altruism," *Quarterly Review of Biology* 46 (1971): 35–57.
9. N. G. Blurton Jones, "Tolerated Theft, Suggestions about the Ecology and Evolution of Sharing, Hoarding and Scrounging," *Social Science Information* 26 (1987): 31–54; H. Kaplan and K. Hill, "Food Sharing among Ache Foragers: Tests of Explanatory Hypotheses," *Current Anthropology* 26 (1985): 223–46.
10. D. Symons, *The Evolution of Human Sexuality* (New York: Oxford University Press, 1979).
11. F. de Waal, *Good Natured: The Origins of Right and Wrong in Humans and Other Animals* (Cambridge, MA: Harvard University Press, 1997).
12. S. D. Preston and F. B. M. de Waal, "The Communication of Emotions and the Possibility of Empathy in Animals," in *Altruism and Altruistic Love: Science, Philosophy, and Religion in Dialogue*, ed. S. G. Post et al. (New York: Oxford University Press, 2002), pp. 284–308.
13. G. G. Gallup, "Self-Awareness in Primates," *American Scientist* 67 (1987): 417–21.
14. G. S. Wilkinson, "Reciprocal Food Sharing in the Vampire Bat," *Nature* 308 (1984): 181–84.

15. J. Alcock, *The Triumph of Sociobiology* (New York: Oxford University Press, 1999); B. J. Bradley, "Levels of Selection, Altruism, and Primate Behavior," *Quarterly Review of Biology* 74 (1999): 171–94; J. Cartwright, *Evolution and Human Behavior* (Cambridge, MA: MIT Press, 2000); L. Dugatkin, *Cheating Monkeys and Citizen Bees: The Nature of Cooperation in Animals and Humans* (New York: Free Press, 1999); E. Sober and D. S. Wilson, *Unto Others: The Evolution and Psychology of Unselfish Behavior* (Cambridge, MA: Harvard University Press, 1998).

16. G. Bernasconi and J. E. Strassman, "Cooperation among Unrelated Individuals: The Ant Foundress Case," *Tree* 14 (1999): 477–82.

17. S. W. Rissing et al., "Foraging Specialization without Relatedness or Dominance among Co-founding Ant Queens," *Nature* 338 (1989): 420–22.

18. M. Shostak, *Nisa: The Life and Words of a !Kung Woman* (Cambridge, MA: Harvard University Press, 1981).

19. Charles Darwin, *The Expression of Emotions in Man and Animals* (London: Murray, 1872).

20. Ibid.

21. G. J. Romanes, *Animal Intelligence* (London: Routledge and Kegan Paul, 1883).

22. J. Archer, "Why Do People Love Their Pets?" *Evolution and Human Behavior* 18 (1997): 237–60.

23. Darwin, *Expression of Emotions in Man and Animals.*

24. Trivers, "Evolution of Reciprocal Altruism," pp. 35–57.

25. De Waal, *Good Natured.*

26. Trivers, "Evolution of Reciprocal Altruism," pp. 35–57.

27. Alcock, *The Triumph of Sociobiology*; Bradley, "Levels of Selection," pp. 171–94; Cartwright, *Evolution and Human Behavior*; Dugatkin, *Cheating Monkeys and Citizen Bees*; Sober and Wilson, *Unto Others.*

28. D. T. Lykken, *The Antisocial Personalities* (Hillsdale, NJ: Lawrence Erlbaum, 1995).

29. Ibid.

30. Shostak, *Nisa.*

31. Blurton Jones, "Tolerated Theft," pp. 31–54; Kaplan and Hill, "Food Sharing among Ache Foragers," pp. 223–46.

32. Symons, *Evolution of Human Sexuality.*

33. Trivers, "Evolution of Reciprocal Altruism," pp. 35–57.

34. P. Ekman, *Telling Lies: Clues to Deceit in the Marketplace, Politics, and Marriage* (New York: Norton, 1997).

35. R. Frank, *Passion within Reason* (New York: Norton, 1989).

36. Ibid.

37. R. G. Simmons, S. D. Klein, and R. L. Simmons, *Gift of Life: The Social and Psychological Impact of Organ Donation* (New York: Wiley, 1977).

38. J. A. Piliavin and P. L. Callero, *Giving Blood: The Development of an Altruistic Identity* (Baltimore, MD: Johns Hopkins University, 1991).

39. Simmons, Klein, and Simmons, *Gift of Life*.

CHAPTER 3: STERILE CASTES OF PRIESTS AND NUNS

1. H. Qirko, "The Institutional Maintenance of Celibacy," *Current Anthropology* 43 (2002): 321–29.

2. Elizabeth Abbott, *A History of Celibacy* (Toronto: HarperCollins, 1999).

3. H. F. Egan, "Celibacy: A Vague Old Cross on Priestly Backs," *National Catholic Reporter* 31 (May 12, 1995): 21.

4. Abbott, *A History of Celibacy*.

5. Ibid.

6. Ibid.

7. G. Boccaccio, *The Decameron*, trans. G. Waldman (New York: Oxford University Press, 1999).

8. L. Betzig, "Medieval Monogamy," *Journal of Family History* 20 (1995): 181–216.

9. Ibid.

10. Ibid.

11. Ibid.

12. Ibid.

13. Ibid.

14. Ibid.

15. Ibid.

16. A. Houtin, *A Married Priest*, trans. J. R. Slattery (1908; repr., Boston: Sherman French, 1910); T. Kselman, "The Perraud Affair: Clergy, Church, and Sexual Politics in Fin-de-Siecle France," *Journal of Modern History* 70 (1998): 587–617; C. Perraud, *Meditations on the Seven Words of Our Lord on the Cross*, trans. St. Joseph's Seminary, Dunwoody, NY (1890; repr., New York: Benziger, 1898).

17. M. Laven, "Sex and Celibacy in Early Modern Venice," *Historical Journal* 44 (2001): 865–88.

18. Betzig, "Medieval Monogamy," pp. 181–216.

19. Ibid.; Egan, "Celibacy: A Vague Old Cross on Priestly Backs," p. 21.

20. Boccaccio, *Decameron*.

21. Donald B. Cozzens, *The Changing Face of the Priesthood* (Collegeville, MN: Liturgical Press, 2000).

22. D. Swenson, "Religious Differences between Married and Celibate Clergy: Does Celibacy Make a Difference?" *Sociology of Religion* 59 (1998): 37–43.

23. M. Dickemann, "The Ecology of Mating Systems in Hypergynous Dowry Societies," *Social Science Information* 18 (1979): 163–95.

24. Marcia Guttentag and Paul F. Secord, *Too Many Women: The Sex Ratio Question* (Beverly Hills, CA: Sage, 1983).

25. Ibid.

26. Ibid.; Abbott, *A History of Celibacy*.

27. Guttentag and Secord, *Too Many Women*.

28. E. Shorter, "Illegitimacy, Sexual Revolution, and Social Change in Modern Europe," in *The Family in History: Interdisciplinary Perspectives*, ed. T. K. Rabb and R. I. Rotberg (New York: Harper Torchbooks, 1975), pp. 48–85.

29. E. J. Sobo and S. Bell, *Celibacy, Culture, and Society: The Anthropology of Sexual Abstinence* (Madison: University of Wisconsin Press, 2001).

30. E. Voland, E. Sigelkow, and C. Engel, "Cost/Benefit Oriented Parental Investment by High Status Families," *Ethology and Sociobiology* 12 (1991): 105–18.

31. Ibid.

32. E. Blackwood, "Culture and Women's Sexualities," *Journal of Social Issues* 56 (2000): 233–38.

CHAPTER 4: WHY DO PEOPLE GROW UP TO BE ALTRUISTIC?

1. C. S. Asa and C. Valdespino, "Canid Reproductive Biology," *American Zoologist* 38 (1998): 251–59; R. Derix et al., "Male and Female Mating Competition in Wolves: Female Suppression vs. Male Intervention," *Behaviour* 127 (1993): 141–74; J. M. Packard et al., "Causes of Reproductive Failure in Two Family Groups of Wolves (*Canis Lupus*)," *Zietschrift für Tierpsychologie* 68 (1985): 24–40.

2. J. Archer, "Why Do People Love Their Pets?" *Evolution and Human Behavior* 18 (1997): 237–60.

3. L. O'Hanlon, "Dogs Outdo Wolves at Reading Humans," *Discovery News*, http://www.wysiwyg://dsc.discovery.com/news/briefs/20030428/dogwolf.html (May 5, 2003).

4. W. P. Anderson, "Pet Ownership and Risk Factors for Cardiovascular Disease," *Medical Journal of Australia* 157 (1992): 298–301.

5. Michael Lewis, *Shame: The Exposed Self* (New York: Free Press, 1992).

6. G. G. Gallup, "Self-Awareness in Primates," *American Scientist* 67 (1987): 417–21; M. Hauser, *Wild Minds: What Animals Really Think* (New York: Henry Holt, 2000); G. H. Schueller, "Hey! Good Looking," *New Scientist* 166 (June 17, 2000): 30.

7. Lewis, *Shame*.

8. Gallup, "Self-Awareness in Primates," pp. 417–21; Hauser, *Wild Minds*; Schueller, "Hey! Good Looking," p. 30.

9. Ibid.

10. F. de Waal, *Chimpanzee Politics* (Baltimore, MD: Johns Hopkins University Press, 1982).

11. Ibid.

12. Gallup, "Self-Awareness in Primates," pp. 417–21; Hauser, *Wild Minds*; Schueller, "Hey! Good Looking," p. 30.

13. A. S. Grutter, "Cleaner Fish Really Do Clean," *Nature* (April 22, 1999): 672–73; L. Shukla, "Cleaning Stations in the Marine World, *Tribune*, http://www.tribuneindia.com (May 5, 2002).

14. H. Davis, "Theoretical Note on the Moral Development of Rats (*Rattus Norvegicus*)," *Journal of Comparative Psychology* 103 (1989): 88–90; H. Davis and S. A. Bradford, "Numerically Restricted Food Intake in the Rat in a Free-Feeding Situation," *Animal Learning & Behavior* 19 (1991): 215–22.

15. R. L. Solomon, "Punishment," *American Psychologist* 19 (1964): 239–53.

16. Asa and Valdespino, "Canid Reproductive Biology," pp. 251–59; Derix et al., "Male and Female Mating Competition," pp. 141–74; Packard et al., "Causes of Reproductive Failure," pp. 24–40.

17. Ibid.

18. R. O. Peterson, "Social Rejection Following Mating of a Subordinate Wolf," *Journal of Mammalogy* 60 (1979): 219–21.

19. F. de Waal, *Good Natured: The Origins of Right and Wrong in Humans and Other Animals* (Cambridge, MA: Harvard University Press, 1997).

20. D. R. Shaffer, *Social and Personality Development*, 3d ed. (Pacific Grove, CA: Brooks Cole, 1994).

21. M. Shostak, *Nisa: The Life and Words of a !Kung Woman* (Cambridge, MA: Harvard University Press, 1981).

22. Shaffer, *Social and Personality Development.*

23. R. Ugurel-Semin, "Moral Behavior and Moral Judgment of Children," *Journal of Abnormal and Social Psychology* 47 (1952): 463–74.

24. Shaffer, *Social and Personality Development.*

25. F. H. Kanfer, E. Stifter, and S. J. Morris, "Self-Control and Altruism: Delay of Gratification for Another," *Child Development* 47 (1981): 51–61.

26. J. R. Harris, *The Nurture Assumption: Why Children Turn Out the Way They Do* (New York: Free Press, 1998); H. Hartshorne and M. A. May, *Studies in the Nature of Character, 1: Studies in Deceit* (New York: Macmillan, 1928); L. Ross and R. E. Nisbett, *The Person and the Situation: Perspectives of Social Psychology* (New York: McGraw-Hill, 1991).

27. Shaffer, *Social and Personality Development.*

28. Ibid.

29. R. Plomin, *Nature and Nurture: An Introduction to Human Behavioral Genetics* (Pacific Grove, CA: Brooks/Cole, 1989); J. P. Rushton et al., "Altruism and Aggression: The Heritability of Individual Differences," *Journal of Personality and Social Psychology* 50 (1986): 1192–98.

30. Plomin, *Nature and Nurture*; Rushton et al., "Altruism and Aggression," pp. 1192–98.

31. Ibid.

32. D. T. Lykken, *The Antisocial Personalities* (Hillsdale, NJ: Lawrence Erlbaum, 1995).

33. N. Barber, *Parenting: Roles, Styles and Outcomes* (Commack, NY: Nova Science, 1998); C. R. Cloninger, "A Systematic Model for Clinical Description and Classification of Personality Variants," *Archives of General Psychiatry* 44 (1987): 573–88; T. Field, *Touch* (Cambridge, MA: MIT Press, 2001).

34. Barber, *Parenting*; Cloninger, "A Systematic Model," *Archives of General Psychiatry* 44 (1987): 573–88; Field, *Touch.*

35. M. Virkunen et al., "Relationship of Psychobiological Variables to Recidivism in Violent Offenders and Impulsive Firesetters," *Archives of General Psychiatry* 46 (1989): 600–603.

36. N. Barber, *Why Parents Matter: Parental Investment and Child Outcomes* (Westport, CT: Bergin & Garvey, 2000).

37. Lykken, *The Antisocial Personalities.*

38. M. L. Hoffman, *Empathy and Moral Development* (Cambridge, MA: Cambridge University Press, 2000).

39. Ibid.

40. D. L. Rosenhahn, "The Natural Socialization of Altruistic Autonomy," in *Altruism and Helping Behavior*, ed. J. Macauley and L. Berkowitz (Orlando, FL: Academic Press, 1970), pp. 251–68.

41. Betty Hart and Todd Risley, *Meaningful Differences in the Everyday Experience of Young American Children* (Baltimore, MD: Paul H. Brookes, 1995).

42. Ibid., p. 177.

43. Barber, *Why Parents Matter*.

44. Beatrice Whiting and John Whiting, *Children of Six Cultures* (Cambridge, MA; Harvard University Press, 1975).

45. Scott Coltrane and Michelle Adams, "Boys and Men in Families," in *The Handbook of Studies on Men and Masculinities*, ed. R. W. Connell, J. Hearn, and M. Kimmel (Thousand Oaks, CA: Sage, in press).

CHAPTER 5: ALTRUISM AMONG THIEVES

1. T. Minogue, "Policing Paradise," *World Press Review* 44 (1997): 42; M. Williams, "Pitcairn: A Two-Century-Old Haven," *New Zealand International Review* 26, no. 2 (2001): 19–23.

2. R. Axelrod, *The Evolution of Cooperation* (New York: Basic, 1984); M. Ridley, *The Origins of Virtue* (New York: Viking, 1996).

3. Ibid.

4. Ibid.

5. Minogue, "Policing Paradise," p. 42; Williams, "Pitcairn," pp. 19–23.

6. Dea Birkett, *Serpent in Paradise* (New York: Anchor, 1997).

7. Ibid.

8. Oscar Newman, *Creating Defensible Space* (Upland, PA: Diane, 1996).

9. "Security Camera Use Is Deterring Crime," *USA Today Magazine* (April 3–4, 1997).

10. N. Barber, *Encyclopedia of Ethics in Science and Technology* (New York: Facts on File, 2002).

11. US Census Bureau, *Statistical Abstract of the United States* (Washington, DC: Author, 2002).

12. E. L. Glaeser and B. Sacerdote, "Why Is There More Crime in Cities?" *Journal of Political Economy* 107 (1999): S225–S258.

13. Ibid.

14. Betty Hart and Todd Risley, *Meaningful Differences in the Everyday Experience of Young American Children* (Baltimore, MD: Paul H. Brookes, 1995).

15. N. Barber, *The Science of Romance* (Amherst, NY: Prometheus Books, 2002).

16. E. J. Sobo and S. Bell, *Celibacy, Culture, and Society: The Anthropology of Sexual Abstinence* (Madison: University of Wisconsin Press, 2001).

17. E. Klungness, "The Lookout," *American Heritage* 48 (1997): 35–36.

18. D. T. Courtwright, "The Drug War's Hidden Toll," *Issues in Science and Technology* 13 (1996): 71–77.

19. J. Vidal, "Doing Borage," *Guardian*, May 26, 2000; "The Garden's Fruit: Prison Life," *Economist*, February 22, 1997, pp. 29–30.

20. John Steinbeck, *The Grapes of Wrath* (New York: Penguin, 1939).

21. Vidal, "Doing Borage"; "The Garden's Fruit," pp. 29–30; John J. Larivee, "Returning Inmates: Closing the Public Safety Gap," *Corrections Compendium* 26 (2001): 1–5.

CHAPTER 6: KINDNESS AND HEALTH

1. Francois Jacob, "Evolution and Tinkering," *Science* 195 (1977): 1161–66.

2. N. Barber, *The Science of Romance* (Amherst, NY: Prometheus Books, 2002); D. C. Geary, *Male, Female: The Evolution of Human Sex Differences* (Washington, DC: American Psychological Association, 1998).

3. R. Insel and T. Hulihan, "A Gender-Specific Mechanism for Pair Bonding: Oxytocin and Partner Preference Formation in Monogamous Voles," *Behavioral Neuroscience* 109 (1995): 782–89.

4. M. Vacek, "High on Fidelity," *American Scientist* 90, no. 3 (2002): 225–26.

5. C. Ezzell, "Brain Receptor Shapes Voles' Family Values," *Science News* 142, no. 1 (1992): 6–7.

6. W. D. Hamilton, "Geometry for the Selfish Herd," *Journal of Theoretical Biology* 31 (1971): 295–311.

7. K. Uvnas-Moberg, "Oxytocin May Mediate the Benefits of Positive Social Interaction and Emotions," *Psychoneuroendocrinology* 23 (1998): 819–35.

8. Barber, *The Science of Romance*; Geary, *Male, Female*.

9. Ibid.

10. T. Field, *Touch* (Cambridge, MA: MIT Press, 2001).

11. Ibid.

12. Ibid.

13. D. T. Lykken, *The Antisocial Personalities* (Hillsdale, NJ: Lawrence Erlbaum, 1995).

14. C. R. Ember and M. Ember, "War, Socialization, and Interpersonal Violence," *Journal of Conflict Resolution* 38 (1994): 620–46; M. A. Straus, D. B. Sugarman, and J. Giles-Sims, "Spanking by Parents and Subsequent Antisocial Behavior of Children," Archives of Pediatrics and Adolescent Medicine 151 (1997): 761–67.

15. Barber, *The Science of Romance;* Geary, *Male, Female.*

16. N. Barber, *Why Parents Matter: Parental Investment and Child Outcomes* (Westport, CT: Bergin & Garvey, 2000).

17. H. F. Harlow, *Learning to Love* (San Fransisco: Albion, 1971).

18. Barber, *Why Parents Matter.*

19. Field, *Touch.*

20. Barber, *Why Parents Matter.*

21. Field, *Touch.*

22. Tiffany Field, "American Adolescents Touch Each Other Less and Are More Aggressive Toward Their Peers as Compared with French Adolescents," *Adolescence* 34 (1999): 753–62; "Violence and Touch Deprivation in Adolescents," *Adolescence* 37 (1999): 735–49; Tiffany Field and M. Diego, "Adolescents' Parent and Peer Relationships," *Adolescence* 37 (2002): 121–30.

23. Barber, *Why Parents Matter.*

24. Field, "American Adolescents Touch Each Other Less," pp. 753–62; "Violence and Touch Deprivation," pp. 735–49; Field and Diego, "Adolescents' Parent and Peer Relationships," pp. 121–30.

25. Barber, *Why Parents Matter.*

26. M. L. Hoffman, *Empathy and Moral Development* (Cambridge, MA: Cambridge University Press, 2000).

27. Barber, *Why Parents Matter;* L. J. Waite and M. Gallagher, *The Case for Marriage* (New York: Doubleday, 2000).

28. Barber, *Why Parents Matter;* Waite and Gallagher, *The Case for Marriage;* J. S. Wallerstein, "Children of Divorce," in *All Our Families: New Policies for a New Generation,* ed. M. A. Mason, A. Skolnick, and S. D. Sugarman (New York: Oxford University Press, 1998), pp. 66–94.

29. M. W. Yogman, D. Kindlon, and F. Earls, "Father Involvement and

Cognitive Behavioral Outcomes of Preterm Infants," *Journal of the American Academy of Child and Adolescent Psychiatry* 34 (1995): 58–66.

30. Barber, *Why Parents Matter.*

31. M. V. Flinn, "Family Environment, Stress, and Health During Childhood," in *Hormones, Health, and Behavior*, ed. C. Panter-Brick and C. Worthman (Cambridge, England: Cambridge University Press, 1999), pp. 105–38.

32. K. Hill and M. Hurtado, *Ache Life History* (New York: Aldine de Gruyter, 1996).

33. H. S. Friedman, J. S. Tucker, and J. S. Schwartz, "Psychosocial and Behavioral Predictors of Longevity: The Aging and Death of the 'Termites,' " *American Psychologist* 50 (1995): 69–78.

34. Barber, *The Science of Romance*; Geary, *Male, Female.*

35. Ibid.

36. Ibid.

37. Will H. Courtenay, "Behavioral Factors Associated with Disease, Injury, and Death, among Men: Evidence and Implications for Prevention," *Journal of Men's Studies* 9 (2000): 81–142.

38. Hill and Hurtado, *Ache Life History.*

39. Barber, *The Science of Romance*; Geary, *Male, Female.*

40. J. J. Lynch, *The Broken Heart: The Medical Consequences of Loneliness* (New York: Basic, 1977); D. Ornish, *Love and Survival: The Scientific Basis for the Healing Power of Intimacy* (New York: Harper Collins, 1997).

41. Uvnas-Moberg, "Oxytocin May Mediate the Benefits," pp. 819–35.

42. Lynch, *The Broken Heart*; Ornish, *Love and Survival.*

43. A. Walsh, *The Science of Love* (Amherst, NY: Prometheus Books, 1991).

44. Barber, *Why Parents Matter.*

45. W. Cannon, "Voodoo Death," *American Anthropologist* 44 (1942): 169–81.

46. R. B. Cialdini, *Influence: Science and Practice*, 2d ed. (Glenview, IL: Scott Foresman, 1988).

47. Lynch, *The Broken Heart*; Ornish, *Love and Survival.*

48. Uvnas-Moberg, "Oxytocin May Mediate the Benefits," pp. 819–35.

49. Ibid.

50. P. Slater, *The Pursuit of Loneliness* (Boston, MA: Beacon, 1970).

51. Lynch, *The Broken Heart*; Ornish, *Love and Survival.*

52. J. J. Lynch, *A Cry Unheard: New Insights into the Medical Consequences of Loneliness* (Baltimore, MD: Bancroft Press, 2000).

53. B. Beit-Hallahmi and M. Argyle, *The Psychology of Religious Behavior, Belief and Experience* (London: Routledge, 1997).

54. C. K. Hadaway, P. L. Marler, and M. Chaves, "What the Polls Don't Show: A Closer Look at U.S. Church Attendance," *American Sociological Review* 58 (1993): 741–52.

55. Lynch, *The Broken Heart*; Ornish, *Love and Survival.*

56. H. Dreher, "Why Did the People of Roseto Live So Long?" *Natural Health* 23, no. 5 (1993): 72–83.

57. Lynch, *The Broken Heart*; Ornish, *Love and Survival.*

CHAPTER 7: KINDNESS AMONG STRANGERS

1. H. Gintis et al., "Explaining Altruistic Behavior in Humans," *Evolution and Human Behavior* 24 (2003): 153–72.

2. Ibid.

3. S. Wilson, *Unto Others: The Evolution and Psychology of Unselfish Behavior* (Cambridge, MA: Harvard University Press, 1998).

4. S. Brosnan and F. de Waal, "Monkeys Reject Unequal Pay," *Nature* 425 (2003): 297–99.

5. A. G. Sanfey et al., "The Neural Basis of Decision-Making in the Ultimatum Game," *Science* 300 (2003): 1755–58.

6. Betty Hart and Todd Risley, *Meaningful Differences in the Everyday Experience of Young American Children* (Baltimore, MD: Paul H. Brookes, 1995).

7. E. Fehr and S. Gachter, "Altruistic Punishment in Humans," *Nature* 415 (2002): 137–40.

8. C. R. Ember and M. Ember, "War, Socialization, and Interpersonal Violence," *Journal of Conflict Resolution* 38 (1994): 620–46.

9. S. J. Emerick, L. R. Foster, and D. T. Campbell, "Risk Factors for Traumatic Infant Death in Oregon," *Pediatrics* 77 (1986): 518–22; R. J. Gelles, "Child Abuse and Violence in Single Parent Families: Parent Absence and Economic Deprivation," *American Journal of Orthopsychiatry* 59 (1989): 492–501.

10. Hart and Risley, *Meaningful Differences.*

11. T. Field, *Touch* (Cambridge, MA: MIT Press, 2001).

12. B. C. Rosen and R. Dandrade, "The Psychological Origins of Achievement Motivation," *Sociometry* 22 (1959): 185–218.

13. N. Barber, *Why Parents Matter: Parental Investment and Child Outcomes* (Westport, CT: Bergin & Garvey, 2000).

14. Ibid.; Field, *Touch*; M. A. Straus, D. B. Sugarman, and J. Giles-Sims, "Spanking by Parents and Subsequent Antisocial Behavior of Children," *Archives of Pediatrics and Adolescent Medicine* 151 (1997): 761–67.

15. Rosen and Dandrade, "The Psychological Origins of Achievement Motivation," pp. 185–218; Barber, *Why Parents Matter*.

16. G. De Vos and H. Wagatsuma, *Japan's Invisible Race* (Berkeley: University of California Press, 1967), p. 23.

17. J. A. Piliavin and P. L. Callero, *Giving Blood: The Development of an Altruistic Identity* (Baltimore, MD: Johns Hopkins University, 1991).

18. A. Kohn, *The Brighter Side of Human Nature: Altruism and Empathy in Everyday Life* (New York: Basic, 1990); B. Schwartz, "Why Altruism Is Impossible . . . and Ubiquitous," *Social Service Review* (September 1993): 314–43; Y. K. Soo and H. Gong-Soog, "Volunteer Participation and Time Commitment by Older Americans," *Family and Consumer Sciences Research Journal* (December 1998): 146–66.

19. Piliavin and Callero, *Giving Blood*.

20. Ibid.

21. "Possible Health Benefits Are Another Reason to Give Blood," *Mayo Clinic Health Letter* 15, no. 12 (1997): 4.

22. Kohn, *The Brighter Side of Human Nature*, ; Schwartz, "Why Altruism Is Impossible," pp. 314–43; Soo and Gong-Soog, "Volunteer Participation and Time Commitment," pp. 146–66; J. K. Murnighan, J. W. Kim, and R. Metzger, "The Volunteer Dilemma," *Administrative Science Quarterly* 38 (1993): 515–39.

23. "Possible Health Benefits," p. 4.

24. J. A. Blake, "Death by Hand Grenade: Altruistic Suicide in Combat," *Suicide and Life-Threatening Behavior* 8 (1978): 46–59.

25. Ibid.; M. Munt, *The Compassionate Beast: What Science Is Discovering about the Humane Side of Humankind* (New York: William Morrow, 1990).

26. S. Oliner and P. Oliner, *The Altruistic Personality: Rescuers of Jews in Nazi Europe* (New York: Free Press, 1988).

27. Ibid., p. 20.

28. Ibid., pp. 227–28.

29. N. Barber, *The Science of Romance* (Amherst, NY: Prometheus Books, 2002).

30. Ibid.

31. L. Malsen, *Wolf Children and the Problem of Human Nature* (New York: Monthly Review Press, 1972).

32. P. G. Hepper, ed., *Kin Recognition* (New York: Cambridge University Press, 1991); R. H. Porter and J. D. Moore, "Human Kin Recognition by Olfactory Cues," *Physiology and Behavior* 27 (1981): 493–95; P. A. Wells, "Kin Recognition in Humans," in *Kin Recognition in Animals*, ed. D. J. C. Fletcher and C. D. Michener (Chichester, UK: Wiley, 1987), pp. 395–413.

33. T. T. Gegax, "A Mysterious Baby Mix-Up," *Newsweek* 132, August 17, 1998, p. 38.

34. R. Jerome, "Growing Pains: 'Switched Baby' Kimberly Mays Fights a New Custody Battle—Over Her Own Son," *Newsweek* 52, September 6, 1999, p. 99.

35. W. Feigelman, "Adopted Adults: Comparisons with Persons Raised in Conventional Families," *Marriage & Family Review* 24 (1997): 199–223.

36. L. D. Borders, J. M. Penny, and F. Portnoy, "Adult Adoptees and Their Friends: Current Functioning and Psychosocial Wellbeing," *Family Relations* 49 (2000): 407–18.

37. G. Slap, E. Goodman, and B. Huang, "Adoption as a Risk Factor for Attempted Suicide During Adolescence," *Pediatrics* 108 (2001): 458.

38. R. H. Frank, "Does Studying Economics Inhibit Cooperation?" *Journal of Economic Perspectives* 7 (1993): 159–71.

39. A. Case, L. I-Fen, and S. McLanahan, "How Hungry Is the Selfish Gene?" *Economic Journal* 110 (2000): 781–804.

40. D. F. Bjorklund and A. D. Pellegrini, *The Origins of Human Nature: Evolutionary Developmental Psychology* (Washington, DC: American Psychological Association, 2002).

41. D. McGinn, "Are You a Tax Chump or a Tax Cheat?" *Newsweek*, April 15, 2002, pp. 38–40.

CHAPTER 8: CONFORMITY AS ALTRUISM

1. M. Argyle, *The Psychology of Social Class* (London: Routledge, 1994).

2. Ibid.; P. Brown and S. Levinson, *Politeness: Some Universals in Language Usage* (Cambridge: Cambridge University Press, 1987).

3. C. Linde, "The Quantitative Study of Communication Success: Politeness and Accidents in Aviation Discourse," *Language in Society* 17 (1988): 375–99.

4. J. W. Grier, *The Biology of Animal Behavior* (St. Louis, MO: Times Mirror/Mosby, 1984).

5. Ibid.
6. N. Barber, *The Science of Romance* (Amherst, NY: Prometheus Books, 2002).
7. Hannah Arendt, *Eichmann in Jerusalem: A Report on the Banality of Evil*, rev. ed. (New York: Penguin, 1977).
8. W. Irons, "Religion as a Hard-to-Fake Sign of Commitment," in *Evolution and the Capacity for Commitment*, ed. R. M. Nesse (New York: Russell Sage Foundation, 2001), pp. 292–309.
9. T. B. Heaton and M. Cornwall, "Religious Group Variation in the Socio-Economic-Status and Family Behaviour of Women," *Journal for the Scientific Study of Religion* 28 (1989): 283–99.
10. B. Beit-Hallahmi and M. Argyle, *The Psychology of Religious Behavior, Belief and Experience* (London: Routledge, 1997).
11. A. Locksley, V. Ortiz, and C. Hepburn, "Social Categorization and Discriminatory Behavior: Extinguishing the Minimal Intergroup Discrimination Effect," *Journal of Personality and Social Psychology* 39 (1980): 773–183; D. M. Messick and D. M. Mackie, "Intergroup Relations," *Annual Review of Psychology* 40 (1989): 51–81; J. Siwqdanius, F. Pratto, and M. Mitchell, "In-Group Identification, Social Dominance Orientation, and Differential Intergroup Social Allocation," *Journal of Social Psychology* 134 (1994): 151–67.
12. H. Tajfel and J. C. Turner, "The Social Identity Theory of Intergroup Behavior," in *The Psychology of Intergroup Relations*, 2d ed., ed. S. Worchel and W. G. Austin (Chicago: Nelson Hall, 1986), pp. 7–24.
13. Barber, *The Science of Romance*.
14. H. Qirko, "The Institutional Maintenance of Celibacy," *Current Anthropology* 43 (2002): 321–29.
15. D. J. C. Fletcher and C. D. Michener, eds., *Kin Recognition in Animals* (Chichester, UK: Wiley, 1987).
16. M. Sherif, *Intergroup Conflict and Cooperation: The Robbers Cave Experiment* (Norman: Institute of Group Relations, University of Oklahoma, 1961).
17. C. Wedekind et al., "MHC-Dependent Mate Preferences in Humans," *Proceedings of the Royal Society* B 260 (1995): 245–49.
18. R. B. Cialdini, *Influence: Science and Practice*, 2d ed. (Glenview, IL: Scott Foresman, 1988); F. W. Young, *Initiation Ceremonies* (New York: Bobbs-Merrill, 1965).
19. Cialdini, *Influence*; Young, *Initiation Ceremonies*.
20. Ibid.

21. Patty Hearst, *People Weekly* 21, March 5, 1984, pp. 76–77.

22. E. Aronson and J. Mills, "The Effects of Severity of Initiation on Liking for a Group," *Journal of Abnormal and Social Psychology* 59 (1959): 177–84.

23. M. Sherif, *The Psychology of Social Norms* (New York: Harper, 1936).

24. S. E. Asch, "Opinions and Social Pressure," *Scientific American* 193 (1955): 31–35.

25. S. Milgram, "Behavioral Study of Obedience," *Journal of Abnormal and Social Psychology* 67 (1963): 371–78.

26. Ibid., p. 377.

27. C. K. Hofling et al., "An Experimental Study of Nurse-Physician Relationships," *Journal of Nervous and Mental Disease* 143 (1966): 171–80.

28. C. Boehm, *Hierarchy in the Forest* (Cambridge, MA: Harvard University Press, 2000).

29. Grier, *The Biology of Animal Behavior.*

30. N. Rothenbuhler, "Behavior Genetics of Nest Cleaning in Honey Bees IV: Responses of F1 and Backcross Generations to Disease-Killed Brood," *American Zoologist* 4 (1964): 111–23.

31. Grier, *The Biology of Animal Behavior.*

32. Barber, *The Science of Romance.*

33. Ibid.

34. P. G. Zimbardo, "Psychology of Imprisonment," *Society* 9, no. 6 (1972): 4–8.

35. Ibid.

36. J. Drinkwater, "Japan: Death from Overwork," *Lancet* (September 5, 1992): 340, 598; R. Lasmont-Brown, "Karoshi—A Fatal Export from Japan," *Contemporary Review* 263 (1993): 197–99; D. Mehri, "Death by Overwork," *International Monitor* 21, no. 6 (2000): 26–28.

37. Drinkwater, "Japan: Death from Overwork," pp. 340, 598; Lasmont-Brown, "Karoshi," pp. 197–99; Mehri, "Death by Overwork," pp. 26–28.

CHAPTER 9: WHEN ALTRUISM FAILS

1. B. Latané and J. M. Darley, *The Unresponsive Bystander: Why Doesn't He Help?* (New York: Appleton-Century-Crofts, 1970).

2. S. S. Brehm and S. M. Kassin, *Social Psychology* (Boston, MA: Houghton-Mifflin, 1990).

3. C. L. Meyer et al., *Mothers Who Kill Their Children: Understanding the Acts of Moms from Susan Smith to the "Prom Mom"* (New York: New York University Press, 2001).

4. D. Symons, *The Evolution of Human Sexuality* (New York: Oxford University Press, 1979).

5. S. R. Dube et al., "Childhood Abuse, Household Dysfunction, and the Risk of Attempted Suicide through the Lifespan: Findings from the Adverse Childhood Experiences Study," *JAMA* 286 (2001): 3089–96.

6. W. Langeland and C. Hartgers, "Child Sexual and Physical Abuse and Alcoholism: A Review," *Journal of Studies on Alcohol* 59 (1998): 336–48.

7. N. Barber, *Why Parents Matter: Parental Investment and Child Outcomes* (Westport, CT: Bergin & Garvey, 2000).

8. Ibid.

9. M. Shostak, *Nisa: The Life and Words of a !Kung Woman* (Cambridge, MA: Harvard University Press, 1981).

10. R. M. Bole and M. Scannapieco, "Prevalence of Child Sexual Abuse: A Corrective Metanalysis," *Social Service Review* 73 (1999): 281–313.

11. H. Y. Swanston et al., "Sexually Abused Children Five Years after Presentation: A Case-Control Study," *Pediatrics* 100 (1997): 600–608.

12. D. M. Fergusson, M. T. Lynskey, and L. J. Horwood, "Prevalence of Sexual Abuse and Factors Associated with Sexual Abuse," *Journal of the American Academy of Child and Adolescent Psychiatry* 35 (1996): 1355–64.

13. N. Barber, *The Science of Romance* (Amherst, NY: Prometheus Books, 2002).

14. Barber, *Why Parents Matter*.

15. Ibid.

16. Barber, *The Science of Romance*.

17. Swanston et al., "Sexually Abused Children Five Years after Presentation," pp. 600–608; Fergusson, Lynskey, and Horwood, "Prevalence of Sexual Abuse," pp. 1355–64.

18. C. M. Cannon, "The Priest Scandal: How Old News at Last Became a Dominant National Story . . . and Why It Took So Long," *American Journalism Review* 24, no. 4 (2002): 18–25.

19. W. N. Grigg, "Wolves in Shepherd's Clothing," *New American* 18 (June 3, 2002): 28–29.

20. Investigative Staff of the *Boston Globe, Betrayal: The Crisis in the Catholic Church* (Boston, MA: Little, Brown, 2002).

21. Vern L. Bullough, "Homosexuality and Catholic Priests," *Free Inquiry* 22, no. 3 (2002): 18–19.

22. K. Shaidle and L. Byfield, "Priestly Sins of the Past: Homosexual Scandals Are Slowly Forcing the Catholic Church to Face Reality," *Alberta Report* 29, no. 6 (March 18, 2002): 49–50.

23. Bullough, "Homosexuality and Catholic Priests," pp. 18–19.

24. A. Ripley, "In Plain Sight: Father Paul Shanley Didn't Hide His Interest in Pedophilia," *Time*, April 22, 2002, pp. 44–46.

25. Investigative Staff of the *Boston Globe*, *Betrayal*.

26. Grigg, "Wolves in Shepherd's Clothing," pp. 28–29; Ripley, "In Plain Sight," pp. 44–46.

27. Centers for Disease Control, "Monitoring Hospital-Acquired Infections to Promote Patient Safety—United States 1990–1999," *Morbidity and Mortality Weekly Report* 49, no. 8 (2000): 149–52; L. T. Kohn, M. S. Donaldson, and J. Corrigan, eds., *To Err Is Human: Building a Safer Health System* (Washington, DC: National Academy Press, 2000); R. P. Wenzel and M. B. Edmond, "The Impact of Hospital-Acquired Bloodstream Infections," *Emerging Infectious Diseases* 7 (2001): 174–77.

28. N. Boyce, "When Healers Kill," *U.S. News & World Report*, January 22, 2000, pp. 46–47; B. Stone, "A Deadly Kind of Care," *Newsweek*, January 12, 1998, p. 33.

29. S. Ramsey, "Audit Further Exposes U.K.'s Worst Serial Killer," *Lancet* (January 13, 2001): 123–25.

30. M. McCarthy, "U.S. Doctor Pleads Guilty to Murdering Patients," *Lancet* (September 16, 2000): 1010; L. O. Prager, "Former Resident Pleads Guilty to Killing Three Patients," *American Medical News* (October 2, 2000): 19; J. B. Stewart, *Blind Eye: The Terrifying Story of a Doctor Who Got Away with Murder* (New York: Touchstone, 2000).

31. J. Adler, "Is California Hospital Worker 'Angel of Death'?" *Washington Post*, February 4, 2001, p. A16.

32. "Death Angel: Ex-Nurse Lynn Majors, Convicted of Killing Six, May Have Murdered Dozens," *People Weekly*, November 8, 1999, pp. 151–56.

33. N. Barber, *Encyclopedia of Ethics in Science and Technology* (New York: Facts on File, 2002).

34. Ibid.

35. R. B. Lee, *The !Kung San: Men, Women, and Work in a Foraging Society* (Cambridge: Cambridge University Press, 1979).

36. D. Humphry and A. Wictett, *The Right to Die* (New York: Harper Collins, 1987).

37. J. Hammer, A. Murr, and D. Foote, "Was Mercy the Motive?" *Newsweek*, April 13, 1998, p. 61.

38. Ramsey, "Audit Further Exposes U.K.'s Worst Serial Killer," pp. 123–25.

39. McCarthy, "U.S. Doctor Pleads Guilty," p. 1010; Prager, "Former Resident Pleads Guilty," p. 19; Stewart, *Blind Eye*.

40. Ibid.

41. Ibid.

42. K. Von Frisch, *Animal Architecture* (New York: Harcourt Brace Jovanovich, 1974).

43. R. Lewin, "The Comeback of Gaia," *New Scientist* (December 14, 1996).

44. L. Jacobson, "Focus on Anger May Help to Curb 'Road Rage,'" *Washington Post*, January 21, 2002, p. A6.

45. K. Patterson, "Experts Note Pattern in Road Rage, Anger behind the Wheel," *Knight-Ridder/Tribune News Service* (July 5, 2002): K7140–45.

46. "Getting the Better of Road Rage," *Africa News Service* (June 21, 2002): 1008172u7449.

47. Jacobson, "Focus on Anger," p. A6; Patterson, "Experts Note Pattern in Road Rage," pp. K7140–45.

48. J. M. Darley and D. C. Batson, "From Jerusalem to Jericho: Situational and Dispositional Variables in Helping Behavior," *Journal of Personality and Social Psychology* 27 (1973): 100–108.

49. G. C. Hardin, "The Tragedy of the Commons," *Science* 162 (1968): 1243–48.

50. M. Ridley, *The Origins of Virtue* (New York: Viking, 1996).

51. J. M. Acheson, *The Lobster Gangs of Maine* (Hanover, NH: University Press of New England, 1988).

CHAPTER 10: TAPPING HUMAN ALTRUISM

1. M. Shostak, *Nisa: The Life and Words of a !Kung Woman* (Cambridge, MA: Harvard University Press, 1981); D. Symons, *The Evolution of Human Sexuality* (New York: Oxford University Press, 1979).

2. J. Alcock, *Animal Behavior: An Evolutionary Approach*, 4th ed. (Sunderland, MA: Sinauer, 1989).

3. A. Jolly, *The Evolution of Primate Behavior*, 2d ed. (New York: Macmillan, 1985).

4. W. H. Durham, "Resource Competition and Human Aggression," *Quarterly Review of Biology* 51 (1976): 385–415.

5. Ibid.

6. N. A. Chagnon, *Yanomamo: The Fierce People* (New York: Holt, Rinehart, and Winston, 1968).

7. N. A. Chagnon, "Life Histories, Blood Revenge and Warfare in a Tribal Population," *Science* 239 (1988): 985–92; L. H. Keeley, *War Before Civilization* (New York: Oxford University Press, 1997).

8. Chagnon, *Yanomamo.*

9. Ibid.

10. Alcock, *Animal Behavior.*

11. Keeley, *War Before Civilization.*

12. Chagnon, "Life Histories," pp. 985–92; Keeley, *War Before Civilization.*

13. Ibid.

14. Ibid.

15. J. Diamond, *Guns, Germs and Steel: The Fates of Human Societies* (New York: W. W. Norton, 1999); J. E. McClellan and H. Dorn, *Science and Technology in World History* (Baltimore, MD: Johns Hopkins University Press, 1999).

16. Ibid.

17. Ibid.

18. D. Brauer, "The Governing Body: Jesse Ventura Grapples with Democracy," *Washington Post*, November 1, 2000, p. C3.

19. T. C. Geononi Jr., "Religious Belief among Scientists Stable for Eighty Years," *Skeptical Inquirer* 21, no. 5 (1997): 13.

20. G. Easterbrook, "Faith Healers," *New Republic* (July 19, 1999): 20–29.

21. H. G. Koenig, M. E. McCullough, and D. B. Larson, *Handbook of Religion and Health* (New York: Oxford University Press, 2001); F. Luskin, "Review of the Effect of Religious and Spiritual Factors on Mortality and Morbidity with a Focus on Cardiovascular and Pulmonary Disease," *Journal of Cardiopulmonary Rehabilitation* 20 (2000): 8–15.

22. Easterbrook, "Faith Healers," pp. 20–29.

23. Koenig, McCullough, and Larson, *Handbook of Religion and Health*; Luskin, "Religious and Spiritual Factors," pp. 8–15.

24. R. A. Hummer et al., "Religious Involvement and U.S. Adult Mortality," *Demography* 36 (1999): 273–85.

25. Easterbrook, "Faith Healers," pp. 20–29.

26. D. Ornish, *Love and Survival: The Scientific Basis for the Healing Power of Intimacy* (New York: Harper Collins, 1997).

27. Easterbrook, "Faith Healers," pp. 20–29.

28. Koenig, McCullough, and Larson, *Handbook of Religion and Health*; Luskin, "Religious and Spiritual Factors," pp. 8–15.

29. R. P. Sloan and E. Bagiella, "Claims about Religious Involvement and Health Outcomes," *Annals of Behavioral Medicine* 24 (2002): 14–21.

30. B. Beit-Hallahmi and M. Argyle, *The Psychology of Religious Behavior, Belief and Experience* (London: Routledge, 1997).

31. B. Clark, "How Religion Impedes Moral Development," *Free Inquiry* 14, no. 3 (1994): 23–25.

32. Ibid.

33. Alfie Kohn, "Do Religious People Help More?" *Psychology Today* 23, no. 12 (1989): 66–67.

34. Diamond, *Guns, Germs and Steel*; McClellan and Dorn, *Science and Technology*.

CHAPTER 11: SAVING THE WORLD

1. R. A. Lovett, "Rain Might Be Leading Carbon Sink Factor," *Science* 296 (June 7, 2002): 1787.

2. G. C. Hardin, "The Tragedy of the Commons," *Science* 162 (1968): 1243–48.

3. N. Barber, *Encyclopedia of Ethics in Science and Technology* (New York: Facts on File, 2002).

4. R. M. White, "Kyoto and Beyond," *Issues in Science and Technology* 14, no. 3 (1998): 59–66.

5. G. Easterbrook, "The Real Evidence for the Greenhouse Effect: Warming Up," *New Republic* (November 8, 1999): 42–63; D. Gantenbein, "The Heat Is On," *Popular Science* 255, no. 2 (1999): 54–59.

6. White, "Kyoto and Beyond," pp. 59–66.

7. N. Barber, *Encyclopedia of Ethics*.

8. P. Barney, "Air Pollution and Asthma: The Dog That Doesn't Always Bark," *Lancet* 353 (1999): 859; P. Cotton, "Best Data Yet Say Air Pollution Kills below Levels Currently Considered Safe," *JAMA* 269 (1993): 3087–88; S. T. Holgate, H. S. Koren, and R. L. Maynard, eds., *Air Pollution and Health* (New York: Academic, 1999).

9. K. Wark, C. F. Warner, and W. T. Davis, *Air Pollution: Its Origin and Control* (Reading, MA: Addison Wesley Longman, 1997).

10. White, "Kyoto and Beyond," pp. 59–66.

11. Central Intelligence Agency, *World Factbook* (Washington, DC: Author, 2000).

12. Ibid.

13. N. Barber, *Encyclopedia of Ethics*; L. H. Newton and C. R. Dillingham, *Watersheds 2: Ten Cases in Environmental Ethics* (Belmont, CA: Wadsworth, 1997).

14. Ibid.

15. Easterbrook, "The Real Evidence," pp. 42–63; Gantenbein, "The Heat Is On," pp. 54–59.

16. Central Intelligence Agency, *World Factbook.*

17. Easterbrook, "The Real Evidence," pp. 42–63; Gantenbein, "The Heat Is On," pp. 54–59.

18. Thorstein Veblen, *The Theory of the Leisure Class* (New York: Macmillan, 1899).

19. C. Oliver, "Why Paychecks Haven't Kept Up," *Investor's Business Daily* 9, no. 23 (1996): 1:1,4.

20. Central Intelligence Agency, *World Factbook.*

21. A. Wilson, J. Uncapher, and L. H. Lovens, *Green Development: Integrating Ecology and Real Estate* (New York: John Wiley, 1998).

22. "Global Warming Alarming? Ask SUV Crowd," *Oilweek* (July 29, 2002): 1–2.

23. "The Global Challenge," *Population Reports* 28, no. 3 (2000): 1.

24. Central Intelligence Agency, *World Factbook.*

25. N. Barber, *Encyclopedia of Ethics.*

26. C. Tudge, *Neanderthals, Bandits, and Farmers: How Agriculture Really Began* (New Haven, CT: Yale University Press, 1998).

27. Ibid.

CHAPTER 12: WHERE HAVE ALL THE VILLAINS GONE?

1. J. C. Maxwell, *The Seventeen Indisputable Laws of Teamwork* (Nashville, TN: Thomas Nelson, 2001).

2. K. Fairbank and J. Landers, "Investigator Says Board, Former CEO Knew Enron Was Hiding Losses," *Knight-Ridder/Tribune Business News* (February 5, 2002), item 02036004.

3. Maxwell, *Seventeen Indisputable Laws.*

4. Ibid.

5. Fairbank and Landers, "Enron Was Hiding Losses," item 02036004.

6. Ibid.

7. J. Landers, "Bush Administration's Energy Plan Included Two Enron-Backed Measures," *Knight Ridder/Tribune News Service* (January 31, 2002): K4039; M. J. Palmer, "Oil and the Bush Administration," *Earth Island Journal* 17, no. 3 (2002): 20–22; A. Park and L. Woollert, "Halliburton: Halfway Home," *Business Week* (December 23, 2002): 54; S. Pizzo, "Family Value$," *Mother Jones* 17, no. 5 (1992): 28–36; David A. Skeel Jr., "The Lessons of Enron," *Books & Culture* 8, no. 3 (2002): 24; A. Zagorin, "An Enron Link to Energy Policy?" *Time*, December 24, 2001, p. 16.

8. N. Barber, *Encyclopedia of Ethics in Science and Technology* (New York: Facts on File, 2002); M. E. P. Seligman, *What You Can Change and What You Can't* (New York: Fawcett Columbine, 1993).

9. A. Raine, *The Psychopathology of Crime* (New York: Academic Press, 1993).

10. C. L. Meyer et al., *Mothers Who Kill Their Children: Understanding the Acts of Moms from Susan Smith to the "Prom Mom"* (New York: New York University Press, 2001).

11. D. T. Lykken, *The Antisocial Personalities* (Hillsdale, NJ: Lawrence Erlbaum, 1995).

12. F. Fessenden, "They Threaten, Seethe and Unhinge, Then Kill in Quantity," *New York Times*, April 9, 2000, pp. 1(N), 1(L).

13. "Live in Fear, Oregon School Killer Is Told," *New York Times*, November 6, 1999, pp. A9(N), A11(L).

14. Fessenden, "They Threaten, Seethe and Unhinge," pp. 1(N), 1(L).

15. Lykken, *The Antisocial Personalities*.

16. Laura Betzig, *Despotism and Differential Reproduction: A Darwinian View of History* (New York: Aldine de Gruyter, 1986).

17. A. Carr, "Is Business Bluffing Ethical?" *Harvard Business Review* (January–February 1968): 143–53.

18. J. Useem, "The Art of Lying," *Fortune* 140 (December 20, 1999): p. 278A.

19. Ibid.

20. Ibid.

21. Ibid.

22. Ibid.

23. Fairbank and Landers, "Enron Was Hiding Losses," item 02036004.

24. "Ex-Enron CEO Skilling Expected to Face Charges," *Knight-Ridder/Tribune Business News* (August 12, 2002), item 0224010; J. Toedtman, "Enron's Fastow Pleads Guilty to Fraud, Agrees to Help Prosecutors," *Knight-Ridder/Tribune Business News* (January 15, 2004), item 04015105.

25. F. Sulloway, *Born to Rebel: Birth Order, Family Dynamics and Creative Lives* (New York: Pantheon, 1996).

26. S. Onyeiu, "Altruism and Economic Development: The Case of the Igbo of South-Eastern Nigeria," *Journal of Socioeconomics* 26 (1997): 407–20.

27. Ibid.

Index